Colour Cash

A holistic design
for an
ecologically sustainable
economic system

Colour Cash

A holistic design
for an
ecologically sustainable
economic system

Barry Voeten

ISBN 978-90-815049-2-8

Original text by Barry Voeten, July 25th, 2006.
Translated by Roeland Peeters and Barry Voeten, 2009.
Published in 2010

Published by Stichting Kleureneconomie,
Van 't Hoffdreef 38, 3146 BR Maassluis, The Netherlands
www.colourcash.org

PDF Book and sources are available on colourcash.org

Soon we must all
face the choice
between what is right
and what is easy

Albus Dumbledore

I. Introduction

Did I need to say that the monetary system is unstable? The Dutch original of this book started off telling you why we really need to get rid of our money system. After the events of October 2008 that seems rather superfluous. By now, everyone knows - or fears it's just a matter of time before the whole thing collapses. Today, October 2009, the Dutch DSB Bank is about to collapse. Small savers have lost confidence in the bank, because, as it appeared, the bank had been abusing customers, selling mortgages at indecent profit rates. The savers withdrew their money and the bank became low on liquid funds. This caused its collapse, although the final outcome is yet uncertain.

This uncertainty is eternal: no one knows the future value of money. We can save money for pensions, but it will always be unsure what the value of that money will be by the time it is paid. Expecting the pension fund to operate fully, of course.

Taken its instability proven, are there any other problems with the money system? Plenty! The monetary system is unfair, illogical, unnatural and unstable. Humanity itself and its money system are both responsible for the horrible state of the planet. Earlier attempts to change the monetary system have been unable to take away its shortcomings. Changes may have had influence, but we've seen no broader effect on society, lifestyle or consumption.

Ecotax - taxations designed to promote ecological activities - has been proposed many times, but it has been implemented only rarely. A powerful ecotax would destabilize the economic house of cards. We've known for many years that we need to change the system in order to save the future, but we cannot without the risk of crashing the whole system instantly.

Considering all these problems I found that a complete redesign of the economic system is required in order to meet social, ecological and economical goals at the same time. Our top-down chain of command requires a change as well.

This book offers an alternative. It is up to you to find out what to do with it. I'm just a designer, not your planet's life-saver. That is something we can only accomplish together! Meanwhile, I'll continue daily life and work in the kitchen garden.

Readers who have not been persuaded after reading the book, do not despair. From Darwin's theory of evolution, we know that none of his opponents has ever walked over to his side. But, his work did have an influence on later generations. They grew up with the new way of thinking. At the time, it took about 100 years for his ideas to be largely agreed upon. Will time fly faster these days?

II. Quick Start Guide

Many users attracted by the subject may find it hard to tune into the spirit of energy and recycling. And it is hard to adapt to a holistic way of thinking. But for some users it may be really easy to tune in. In the past we have recognized the following groups of future users and their ability of a quick start:

- Artists and designers. Most often used to not think top-down and not to count money as a measure of result. Intuitive understanding of the colours. Please proceed to page 58.

- Users of the permaculture garden / lifestyle approach. The hands-on counterpart of colourcash. What's new here is a way to count your blessings in numbers and possibly, a money system that suits permaculture. Please proceed to page 58.

- Cradle- to-Cradle businesses. Already on the path of closing life cycles and reducing energy supply. Where Cradle-to- is a way to introduce ecology into business, colourcash may be a way to introduce business into ecology. Please proceed to page 58.

- Transition Town participants. An approach to reduce dependency on oil and prepare for climate change and peak oil. I hope this book may take your work to another level.

III. Outline of this book

In the first part, we go through my personal history. Not that I am that important, but we will take the steps that I took to start this design. Many things will sound familiar, probably. Also, it serves as a shock therapy, getting those that didn't use the quick start ready for the next level. An important document included is the Enschede Vision, a document a group of students wrote for the benefit of the future of the City of Enschede in The Netherlands. This vision shows where we should go: a sustainable city in a sustainable world. The rest is just a vehicle to get there.

The second part contains the nucleus of colourcash Trading. This includes both the new monetary unit of value and the new rules of the game. The economic chain of production will be replaced with an alternative from biology: the life cycle. The spirit of fight and competition makes place for the cooperation. A clear definition of the unit of value is presented, following the examples of the definitions of distance, weight and time. This sums up the contents of part two.

In part three we will look at our neighbours: existing alternatives and changes to the monetary system. This part is rather short regarding the available material, but we think we get the idea. In the upcoming book, other related subjects such as cradle to cradle, permaculture and transition towns will be presented, but in this book, we are unfortunately doing without.

In the fourth and last part we will consider the consequences of colourcash on social and economical topics. What happens to taxation, unemployment, international trade? What are the similarities and the differences. These kind of subjects are discussed from A to Z in part 4.

What it's all about when it comes to colourcash? It's about you! This book is not just to keep you busy on rainy Sunday afternoons. This is about recovering from a collective brain malfunction: the strive for continuous economic growth in a world of competition, struggle and fear. Let's get ready for a world of cooperation, love and fun! Enjoy!

Barry Voeten

Maassluis, The Netherlands. January 2010

Table of Contents

Part 1. Origin

1.1 An obstacle to development

I wanted to write this book a long time ago, but I never got round to it. I thought I didn't have the time... This is what happens when you have "a job". You have to make ends meet, make money for "every-day" needs, to do "fun stuff". To earn all this money I too have to work. So you don't have time to do what you wanted to do years ago.

A friend of mine has his own company. Often working till 7 or 8 pm and then off to the tennis court. One day he was at the tennis club with his wife and son. His son had entered the court with street shoes, something that was and is prohibited. His father was proud. I could tell he thought it was great to see his son grow up. And that he, as a matter of fact, had too little time to see his child grow up.

One sunny afternoon he collapsed in the middle of town. The hospital doctor diagnosed epilepsy. And it was time to down shift. De car was traded in for a bicycle, the tennis court for the living room. I only saw him on the odd occasion after that, but I do know he's at least just as happy as before. A lower gear, a gentler speed.

I've been working in the Internet industry for years. The rise of this medium, somewhere at the end of the nineties, was explained very simple in the industry: 30% of Internet traffic is pornography. The expensive connections wouldn't be profitable if it weren't for the demand of pornography.

What about the supply? Considering the number of models on offer, you would almost be amazed to find a woman who hasn't sold her body yet. The origin of this enigma is easily explained. You can easily see that the hotel rooms are in another part of the world, another continent, by the cheap wallpaper. Lots of girls from Eastern Europe, Ukraine and Biella Russia find their way, often digitally, to the Western-European and U.S. sex industries.

This without mentioning the sex tourism. Thousands of horny men go to the Far East to be "served" by children. And, closer to home, in Rotterdam, they just closed de Keileweg, noted area for drugs prostitution.

Drugs are a plague for the mind. Often these drugs are cultivated by poor farmers in South America, smuggled by people who don't have any chances up to their ears in debt who spend many long years behind bars when arrested. And the Mafia gets very very rich...

If you aren't satisfied yet by all this bad news, here's just one more story. The mill here in Maassluis, a small city in the industrial area of Rotterdam , sells delicious cake-meal. Even though the stuff contains a lot of sugar, our family uses it on regular bases. The woman who sells the flour told me that it all comes from our the neighbouring town, Vlaardingen. The mill in Maassluis can't be used. That is to say, it can turn, but can't be used for milling. They have added so many rules and regulations; it's just no fun any more to mill, the woman told me. The wheel has been declared "unfit". A shame really, because the mill is a windmill, it works when there is wind... for free. But it turns idle. Pure waste.

There are many branches in the industry that suffer from an abundance of rules and regulations. And what is the use of all these rules? To guide the economic process, to prevent usurers and clumsy oafs from running mills they can't operate. Because our daily bread could contain little stones so the people might have to go to the dentist because there was a little piece of stone in their bread. Instead the rules and regulations created mills that turn but don't work.

Every one of these anecdotes has the same, typically human motivation. Making money. If there was no money to be made, the people in the examples would have made other choices, or rather, they would have had another option. If people would not follow the path of money, politicians and lawmakers would have made other choices. So we make the world to make money of it. As a result we live to make money and everything has to make way to make money as well.

In my own life it's not different. The things you do for money's sake, you'd prefer not to do if you weren't paid to do them. This doesn't mean you're good or bad, or that money is good or bad, because it just isn't that simple. I'm only trying to illustrate that, what we consider the most normal thing in the world, in fact has an abnormal cause.

What leads us to place such a high value on money? What intrinsic value does money have? I will try to answer some of these questions in the next chapters. Or rather, I will do my utmost to kick the last cripple leg from under the monetary system... trust. Even though some of my friends told me the Euro already did that. The Euro comes in handy when we go on holiday, but other than that we have been fooled.

Whether it's called Euro, guilder or neo-capitalist market economy, I don't trust it very much. To show you exactly WHY I don't trust it, I will take you on a trip through my life, a cruise down memory lane if you will. You will also see why I propose an alternative option. It's easy enough to moan and groan, but to propose a solution is another matter. It has taken me 33 years.

1.2 Child of the times

When I was a boy I had a strong intuition. My only mistake was thinking I was smart. But it was only the contact with my own "inner self" that gave me the answers. Many years later I would find that out. For the time being I am still in primary school and my class mates call me "professor". I was obnoxious and I used to think to have the right answer to every question. This is irritating, whether the answer is correct or not. I haven't changed much in this respect, but you will find that out soon enough.

I changed school when I was in third grade. From now on I went to a Montessori school. There I could set my own, rather fast pace with the school subjects and I didn't have to wait for the slower students any longer. This school is where I have my first memory about money. I was used to listening to myself to find answers. And especially in arithmetic I was a star. There were other, smarter kids. There were always other boys or girls, even from another group who had higher marks with greater ease. I got the answers by listening to myself. Half a second is / was enough to know the answer.

In the same intuitive manner we had a conversation with a group of children about adults. It was around the year 1984 and I think it was Prime Minister Ruud Lubbers who implemented some drastic measures.

We children came to the simple conclusion:

Adults do anything for money
and money only.

We thought you couldn't eat money and that the world was a lot more important than money. However, in the eighties, the world was changing. It was the era of the Young Urban Professionals, the era in which people loved getting rich, the epoch of color TV with their nice girls, big cars and expensive suits. The arrival of the CD player was a major revolution and brought us into the so-called digital era. We had a computer at school, a Commodore 64. The thing had hardly any memory, not even one thousandth of an mp3-player kids use nowadays. To load a game you had to insert a cassette which transmitted a code by means of an endless series of beeps to a brown case we called "the bread box". Al these technological developments came into being through enormous investments from powerful stockholders.

Nonetheless, those early pioneers tinkered lonely and not understood in their garages with primitive equipment. Only when there was real money to be made , computers took off. And now, you and I and the whole world are connected via cables and wires.

Even then, at the age of 11, I had an image of myself :

One day, right now I don't know

what exactly,

I'll do something really important

with that money

It was a very strong sensation after we discovered that adults are only occupied with money and destroyed everything for money. When I see a picture of myself from then, I see a self righteous punk that had all the answers, even more so than me years later.

Some years later, in the eighties, I went to high school. First in Amersfoort in the centre of the Netherlands, but two years later we moved to the South of the country, where my crib had been. All this had to do with my father's job.

My father worked as an accountant for a record company which was part of the huge Philips-concern. Philips had to make records as well, in order to ensure the sale of turn tables, and that is why Philips had a record-emporium.

My father worked as supervisor on a division which had to do with money, accounting. My mother also worked in accounting. My sister and I passed by her office quite often, as it was near to our school. You could see how numbers were added horizontally and vertically. When the totals of the totals matched, the scheme was right and another page was booked. It looked difficult, but I got the idea.

Anyway, my father had been working for this record company for some years and things weren't exactly going great. Of course, he was just as hard headed as I was, and so he wasn't mister popular with his superiors. Looking for another job he ended up in a construction company in the southern part of Holland. For one year he commuted daily about 150 kilometres by car, before we moved house.

Being the son of two economists I spent my puberty in a small village in the south. I was the only one in the household with a knack for technology. Even though my sister had a better grasp of things like, e.g. electricity than both my parents put together, I remember that I always had to figure out the handy work myself. Compared to other boys who had a handy man for a father and the necessary tools, I was almost crippled. My sister, two years older than I am, eventually chose the familiar path of economics. Luckily she had some brochures from the TU in Eindhoven, and that is where I got my inspiration. Economics were something for wussies I thought and I wanted to study something really technical. Where else could I go with my knack for maths? OK, I was one of the few students who had taken the subjects "economics 1 and 2", but this had more to do with the fact I thought these subjects were easier than

Figure 1: Bryan Romero's idea of worthless money.

anything else. I got the jargon fed into me with my mother's milk, as it were. The typical economical thinking wasn't a problem for me, even though I had some questions about the vagueness of economic theories.

There was, for instance, that story about inflation. One of my mates and I thought this inflation was a load of crap. How could money lose its value from one day to the next?

We had an answer to this: It was this easy; Money has no intrinsic value. It's only worth what the fool will pay for it (breads, milk, a house...)

One day people are a bit less crazy and find out that money is worth less and milk a bit more. We split our sides with all the silliness of the economic system and the amateurism they tried to describe their models. And our teacher said:

These are simplifications because the real models are way too complicated to teach in this school.

We weren't impressed, because in Mathematics and Physics we had to deal with matters that were a lot more complicated, like sinuses, exponential functions and calculus. You could say a lot more with those...

My neighbour and I had the following image. These economists are just unlucky so-and-so's for whom real mathematics are just too difficult and therefore went to study economics. And this poor teacher couldn't get a real job in the industry, she didn't know how to divide 15 by three. She didn't know whether you had to put the 15 on top and the 3 below or the other way round.

As it turned out there was a "real" and serious branch within economics, a section where they could do the maths. Econometrics it was called. The boys and girls end up in clubs where they are used as assistants to calculate political models. In fact this is even worse: you know your maths and economics and you let yourself be dictated by political ignoramuses who barely know anything about either one. You only went to study law if you couldn't do maths or economics, I thought. And that is why lawyers are the ones who complicate life for the rest of us so much. They tie us to a set of rules that, together with the pressure of the bank, can make life really hard.

I have to admit that back then I hadn't woken up to the fact that all those people who thought up those rules and took those short sighted decisions only did it because they had no choice in the matter. They too have wives, children and mortgages. So, who is it that can afford to shout "No, That never!" and bang their fist on the table? Correct. The unemployed and people who are on the outside of society. And no, not even them, because looking for a job is a full time occupation these days with the same pressures as a real day-job. You have to enrol everywhere, fill out an enormous amount of forms, show them that you have a bank account... All this before you see one cent! I was employed again by that time... and had been for a long time.

But I was one of the more fortunate that could afford to study. My parents had budgeted wisely and my father had a nice income and with that he could afford to have us study. This was important for him, because he never had had the chance to do the same.

Grandpa was a factory worker. Plodding on, early up every morning, heavy labour in the rubber industry... Dangerous work at that. It cost him half his hand. The work had to go on. Moonlighting in gardens... The three sons were bright enough, but not a hope to send them to university. If you wanted to study, you had to work your way through college. I don't really know how my uncles did it, but father became an accountant. So a full time job from the time he was 18 and study at night for about ten years until you're done.

Meanwhile you get children. No real choice, little time, because the children have to be able to study when they grow up, give them the choice he never had. A noble goal that nevertheless wasn't always understood. Because when you're 5 you go to bed at half past seven. And daddy has only been home for about an hour at that time. Daddy didn't have the choice or he had already made his choices.

I myself was able to go to college and I chose IT. I didn't know anything about it, but that didn't matter, that is exactly why you study, to learn... It was the year 1990 by now and some families had a computer at home. My father had thought me that those weren't the "real computers" but how and why... I didn't have a clue.

I chose IT for one reason really... It was somewhere between maths, economics and electronics and you could always decide which direction to take later on. At least that was the official reason I gave when asked. Inside of me I only knew it had something to do with computers, and computers were the novelty of the time. It was the only reason I ever needed.

It's strange how the most important decisions in life are taken on a whim. This would happen to me many a time. Girlfriends, houses, jobs... I don't have it in me to really think those things through. I only think up reasons afterwards to apologize as it were for a decision I took on a whim.

And so I ended up in the University of Twente, because it has a nice park with student lodgings close to the auditoriums. I had seen the University of Eindhoven but I didn't like it because it was a concrete giant in the heart of town. Delft surely wasn't anything but concrete as well, and years later it turned out that I was right. My friends went to Eindhoven because it was closer to home. This way they could still live at home the first year and later they would decide what to do. In case university didn't work out they could still lower their aim and go to HTS (kind of technical university/college). I rented a room straight away to get away from my mother's skirts as soon as possible.

1.3 The Academy Unmasked

I was set to go study real science, real technology that was "about" something concrete. We got some pretty hard maths to chew, but we were spared from the worst compared to the people in Technical Sciences. Always "one up" of course. One of the most remarkable things in maths is that complex, unintelligible, and unexplainable things can be set in a formula with which you can do the most fantastical things.

Also in IT we use mathematical descriptions to base our program code. In fact, a computer is nothing more than a complex calculator. It's the logical result of perfectly programmed functions. Even if your computer crashes, it's the result of a phenomenon than can be described mathematically. Unfortunately the programmer didn't take your situation into account and so the program got out of it's frame, so the space next to the program was affected, and then the damages where substantial, nothing reacted, except the on/off switch. This same principle, the "crash", can be applied to a real person or country when they go bankrupt. They have to start from scratch again.

In the IT course you dealt with Plato's Technical-Scientific school of thought up to WWII, leaving aside biology and anatomy. There were some interesting parts, I must admit. Even though there was a trend that you were taught a lot about something you didn't know the use of. If you were able to perform your tricks on your finals, you scored another subject. After little over a year I didn't see the use of this and set myself another task: exploring the world and the discovery of sex, drugs and Rock 'n Roll. Of course these discoveries where joined by philosophizing and the "What am I doing on this Earth" question. One of these discoveries was the new (read:1992) report by the Club of Rome.

I was reading a newspaper in the kitchen of the student flat where I was living at the time in Enschede. The news was old but horrendous. We can't go on like this. Society, oil, population, food... everything was caught in an exponential growth.

Exponential growth is a lot like a flea infestation. A rabbit farm. It boils down to this: every so often, e.g. a day, the number doubles. For instance, today 1, tomorrow 2, the day after 4, next week 128, next month 2147483648.

The only thing that does stop exponential growth is negative exponential growth. And that is what's happening now. The pressure on oil wells is slowly decreasing, oil production is stagnating. In the back of the book, under the Subject Peak Oil, we'll come back to this.

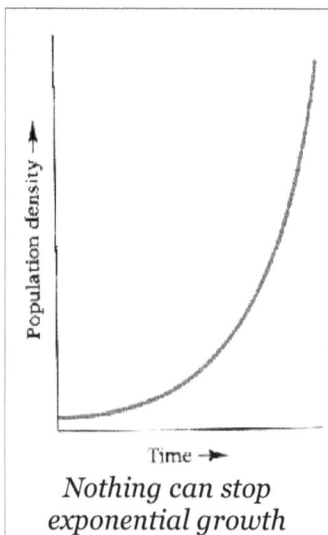

Nothing can stop exponential growth

In so-called "limit-mathematics" this is standard material for high school students. I knew it, my classmates knew I, the whole Technical University knew it. But nobody was going to take any action to the "Club of Rome" report. While we, as far as I could see, were the only ones who could do see that there was something seriously wrong. Nobody seemed to care whether we would live or die, as long as we see tomorrow or next week.

The media sustained that it wasn't all that bad. The Club of Rome had been mistaken, because there were a lot more supplies than the 1972 estimates predicted. The newspapers didn't realize that the model still was in place and that exponential growth had been reached. Maybe it was a trick of the media to leave people in the dark and distract attention.

Then I decided this:

If nobody does anything about it,
I will.

Thirteen years later I still think the same way. Meanwhile a lot has been done, but few goals have been reached. Nevertheless, the world has changed a bit. But first I will tell you how things were at university, because that explains a lot about the present.

I was living in this abstract mathematical dream world that can be described perfectly using equations. The most important equation at University is not science, but budget. Instead of working together, there is the eternal struggle for money, projects, positions, connections.

I ended up in the research commission via one of these connections. This commission had as main task to advise the Faculty board about research. Back then there were two positions for students, and about 5 professors or teachers. As a student-member you got a nice remunition to go through the papers and to attend the monthly meetings.

During the two years I was a member of this commission, not one of the other members put a question forward about research. Never. The only questions were about FTE's, an abbreviation for the number of jobs available, or, in other words, money. Are these FTE's justifiable? Right.

One of the members –and at the same time one of my teachers- told me proudly that his position at the University was only due to the publication of an article in the same magazine, in the same edition the professor had published. Both article were against a competing research area, AI, Artificial Intelligence. AI was found too "vague" by technicians. Excellent reason for a position, isn't it !?

I saw that the main objective in science was "competition". Science itself had taken a back seat, after the struggle for power and money. Nowadays, fortunately, co-operating is becoming more and more the rule. Has to be. A number of research foundations has been set up to join forces from related areas of knowledge. That group gets funding and competes only against similar groups elsewhere in the academic world. Or, the battlefield has moved and now we form alliances against similar institutions elsewhere in the academic world, against another collective.

Rock hard technology and beautiful science
had succumbed and
had become slaves to the economy.

This insight still explains why I don't allow researchers to seduce me. And it also explains why they don't like me, because the way I work doesn't give them any financial gain. It's only logical that those who have sold out to their mortgages no longer are able to come up with a creative solution that could put that same mortgage in peril. And this is true for anyone. If you are up to you ears in debt, loaded with money or still believe zealously in the "money is important" illusion, this story will be hard to digest.

The remainder of this book is only for those that have woken up to the idea that the whole "money-idea" is nothing more than an illusion. We will provide you with another illusion, then it's up to you to choose which illusion you want to live.

1.4 Introduction to The Enschede Vision

Right, I was set to, even if nobody else would, save the world. I became active in the Enschede scene. I joined a youth club for Sustainable Development. A real buzz word from the early nineties. What did sustainable mean again back then?

Mrs. Brundtland from Norway wrote a nice report (Our Common Future), I quote:

Sustainable development is to
fulfil our present needs
without harming future generations
in the fulfilment of theirs.

Soon I had another definition:

Live in such a way you can sustain it,
you can keep it up.

It was an unusual club because we really worked together. In spite of the fact there were kids from various religious and political views. We were there for a common goal. The Jongerenplatform Duurzame Ontwikkeling Enschede (JP Doe) had as its goal to make and keep the city liveable in fifty years time. So far it's been 10 years, still 40 more to go.

One of our first actions was a successful lobby with the town of Enschede to join Local Agenda 21. That's a story in its own right. In 1992 the United Nations decided in Rio de Janeiro to "do something about it", on the UNCTED conference. They too had read the report of the Club of Rome. But they decided they couldn't do anything about it! We, world leaders, decide here that we won't do anything about this problem because we are not able. We are too far away from the citizens.

That's why they called upon the communities of the World to join the Local Agenda 21 (LA 21). Let the municipalities figure it out. And so it was. Thanks to the VOGM subsidy of the Ministry the LA 21 was introduced in several Dutch municipalities. In Enschede we got a grant of 200.000 Guilders. That was our first feat of arms.

But it gave us some considerable problems: now the Municipality is joining in, what do we want them to do? Thanks to the amount of action material that we received we discovered the Vision-Strategy-Action point approach and decided to form our vision first. The Enschede Vision.

About ten youths met up every two weeks, discussed, made notes. Those notes were elaborated and discussed again. After almost 18 months it all became clear. Everything, absolutely everything has to be brought back in a cycle and those cycles have to be kept as small as possible, by using as little energy as possible. It's the only way to keep everything going.

This still is the foundation of colourcash. These are the principles that make the economy, from a technical point of view, work. The rest is a game, a modus operandi of cycles and energy. The principles themselves are best explained in the original Enschede Vision.

1.5 The Enschede Vision

Local Agenda 21 seen for the energy prospective

1.5.1 Introduction

This vision is seen from the perspective of energy. The perspective has a natural principle.

The law of conservation of energy.

Every process uses up the same amount of energy it produces. The energy income always originally derives from the sun.

A big part of the energy is transformed in heat that gets lost in space. The energy reserve available on earth, in the shape of fossil fuel, is limited. Also then uninterrupted solar energy supply has a limit.

In the rest of this chapter we will discuss the 7 themes of LA 21. In order to avoid confusion the first theme of LA 21 (International) is dealt with last. We also added an extra theme, water.

1.5.2 Theme 2: Energy

As it happens the theme and the perspective are the same.

Natural energy production

All energy eventually originates from the sun. The best way to go about Sustainable development is to contain, to keep this energy as long as possible in a usable form before it disappears into space. This means that we need as much green (nature) as possible because plants are the best sunlight to energy converter.

Plants are food for mankind and animals alike.

So, indirectly, people eat sunlight.

The amount of energy used shouldn't be higher than the average caption of solar energy. This principle has far-reaching consequences for the following themes.

Technical energy production

Mining rather harvesting fuel mustn't impede natural energy production (hydro-electricity and strip-mining). These types of exploitation do not permit plants to collect solar energy. The plants are no longer edible. This way so-called energy production by humans" impedes the energy production and storage by nature.

We must strive to a cleaner way to convert energy. With the notion Clean Energy we mean the energy conversion processes whereby the environment is not affected by undesired side products (CO_2, SO_2, NO_x, acid rain).

During the life cycle of the energy converter the energy use mustn't be higher than the energy-production cost. A solar cell costs (read: in 1995) the same amount of energy to make than it can produce during its useful lifetime. So a solar cell actually costs energy because we haven't even considered the price yet. You can see similar paradoxes with e.g. modern (electrical) windmills and so-called energy efficient lamps.

Conclusion

The only clean energy based on present technology is agriculture, forestry and fishery, if done well. This means it can not be slash and burn or predatory exploitation. This form of clean energy should be sufficient.

1.5.3. Theme 3: Building and Living

Building takes up a lot of space which could have been used for energy capture by nature. "Space-efficient per person" living gives us energy because you conserve it instead of transforming it into heat. (Have you ever walked across a full parking lot in summer??) Building, heating, maintaining and demolishing buildings costs energy.

Newly built neighbourhoods put more pressure on the environment because of the need for more roads. When planning building and living, minimizing the energy to be invested must be taken into account.

The energy-cost of demolishing a building has to be taken into account in its design to prevent waste of both material and energy.

In architecture the flexibility in use of the building has to be considered. This means that buildings must have a multi-functional character, so you can use them no matter what the application.

Life styles in the sense of single, family, commune, influence houses and abodes. Group housing with common and private spaces diminishes the number of kitchens and sanitary provisions needed. The energy needs of group housing is lower in principle.

1.5.4. Theme 4: Waste and raw materials

In nature all processes and materials are organized in

cycles. We must strive to get as close as possible to these natural cycles. Nature shows us that all processes are closely interwoven in Dynamic-cyclic balances. Cultural processes can only continue to exist if they are in dynamic-cyclic balance as well.

All production and demolition processes use energy. It's a matter of life or death to produce in such a way that the cyclic production, use and destruction demand as little energy as possible. The waste-phase is the transformation of products into raw material or half-products. Also this requires energy and has to be taken into account in the price of the product.

To get rid of waste you can burn it. That way you can use the last bit of energy. The result is a heap of rubbish you can't get rid off. This rest is very energy and time expensive for nature to incorporate it into a cycle. You can't keep this up indefinitely and so it is undesirable in principle.

So the answer here is to try to clean up the waste deposits and re-use as much material as possible. Alas we must accept loss of material for existing waste deposits. Not all garbage is re-usable, but a lot of waste is. Future dumps are avoidable.

1.5.5 Theme 5: Traffic and Transport

Moving people and goods uses up a lot of energy and seriously pollutes the air. We think that the traffic-need can be reduced drastically, but we need to change our "mobility behaviour".

In the transport of goods a decrease of kilometres travelled, and so used energy, can be brought about by developing a consumption behaviour which is based on the consumption of local goods.

Transport of people in commuting is not energy efficient when a car is used. The energy the car uses to get to the pace of work is a lot higher than the energy invested in the actual work. If the distance between the home and the place of work is reduced, this can help save energy. The need to use a car disappears.

In towns and cities we must create possibilities that make the bicycle the fastest mode of transport. The space used by cars should be reserved for bicycles. Table 1 shows the various means of transport and their energy use.

Vehicle	Fuel used per 1000 km p.p. in litres
aircraft	100 l
car	67 l
bus	26 l
train	17 l
bicycle	nil

Table 1: Vehicles and fuel usage. NJMO & IPP, Future in Action, 1994

1.5.6. Theme 6: Agriculture and Food

In Agriculture solar energy is transformed into a form that can be used for human consumption. Maximum profit based on an infinite time scale is vital. Therefore soil exhaustion has to be avoided. Food and fibre production has to be aligned with local consumption of humans and livestock.

Organic agriculture has to be the standard to follow. Farmers and consumers should develop the insight that it's more expensive to pollute the surface water using fertilizer and pesticides. If surface water is kept clean, we save energy which is needed to render the environment optimal for man and his surroundings. The means and methods used in agriculture are not energy efficient, they use more energy than they yield.

Local, "green" produced products are sold in all possible sales points, and therefore also in supermarkets. Packaging food can't use more raw materials than strictly necessary.

Food is the end product of a cycle, clothes, however, aren't. It's possible to change the mentality in buying new clothes. The latest fashion needn't be the main reason to buy new clothes. After clothes have been worn they can often be used again. There are numerous social institutions that would be very happy with used clothing. This form of donations should be stimulated more. Worn clothes still contain fibres that could be re-used in the production of new materials.

One alternative for choosing, say a new pair of trousers, is looking for durable fibres, fibres that have a higher wear-and-tear resistance than the fibres now commonly used. One of those fibres is the Cannabis, or rather hemp plant.

Intensive agriculture should be transformed into extensive agriculture and agriculture should be on a smaller scale with mixed produce as opposed to monoculture. Research has shown that cattle farms which have greater biodiversity have the advantage when confronted with a disease in the livestock.

1.5.7. Theme 7: Nature and Scenery

Nature and scenery are the surroundings of man and is indispensable for the survival of all living things. Small natural cycles, the various cycles of life, make up the big cycle of Nature and that is why it's desirable to connect every level within these cycles. A maximum biodiversity is needed. It is important for a firm natural balance. The opposite of this is the desire for maximum profit. This leads to soil exhaustion and interrupts various natural cycles.

Green, especially trees, have the highest yield as converters of solar energy to stored energy. A greater biodiversity can be reached by planting primary forests. This form of forestry leaves more possibilities for the storage of big quantities of energy than planted wood fields, production woods.

1.5.8. Theme 8: Water

The most important and most expensive raw material is water. It takes a huge amount of energy to desalinate seawater for use as drinking water. The sun supplies drinking water by means of clouds, prevent acid rain by avoiding exhaust gases! We must not extract more clean water from the soil than the amount of clean rainwater that goes into it. Agriculture and heavy industry often waste enormous amounts of water, and this becomes apparent in the price of their products. They also have to provide purification for the water they use. Consumers can contribute by recycling water used for the shower to flush the toilet using a secondary water circuit.

1.5.9. Theme 1: International

To implement all the local themes dealt with so far on a global plan, we must strive towards local economic regions that exchange products on a small scale amongst themselves. The environment should be the starting point for the set up of this economic infrastructure and not as final goal. The continuation of large scale exchange between globally aimed structures (multi national companies) should be reduced. This implies a reduction in then world wide transport where similar products are flown back and forth in the guise of economic progress. Unnecessary transport of goods and people is a pure waste of energy.

So here a complete chapter from the Enschede Vision. Now you know where we're going. It's our goal to really implement all these principles. The question remains: How?

The remainder of this book is about looking for a way to implement this vision. We still have a long way to go. We came to the conclusion that it's a matter of "making the impossible possible" and not so much a matter of making all the necessary changes ourselves. In the end people have to do it themselves.

The next point on the agenda after Vision is Strategy. How do we come to a good and useful strategy? I needed to follow a few more courses at University first.

1.6 Looking for a meaning

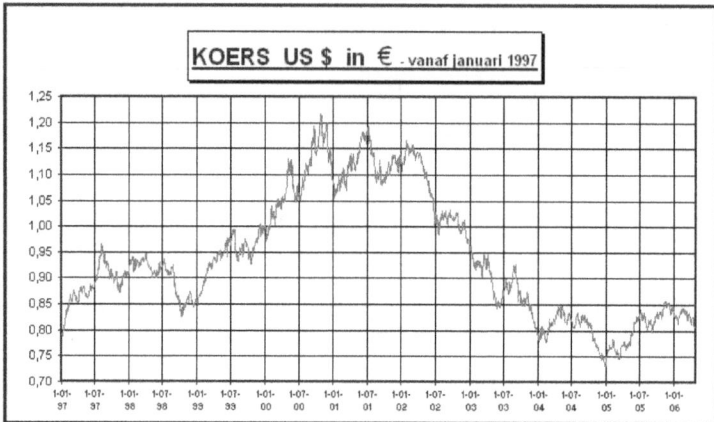

The "size" of the Dollar changes continuously. How can you use a changing unit as a foundation for a society?

One of the courses was about the semantics of programming languages. What is meant by this is; the formal meaning, solidified rock hard. An example. Everybody knows what a meter is. The real, original meter lies heavily guarded in Paris, because that measure was the calibre for all the copied meters. A clear example of semantics: record precisely what you mean. Write down exactly what you mean: the meaning.

Meanwhile I was still stuck with the Enschede Vision and the economic reality; even science had succumbed to economic "progress".

Then I found out the secret of money: it has no meaning! No meaning has ever been set. The meaning of money as such is based on trust, something I can't have in money. The days that

32

money could be exchanged for its value in silver or gold are over, and have been for a few decades.

When you ask Google "define money", you get a number of articles on how money works. Money is a means of storage and exchange of value. How this value is determined is unknown. Compare to this: a meter is a unit of length. How long the meter is, is actually unknown. So, 1 dollar equals 100 dollar cent. Genius. One meter equals 100 centimetres. And how long is a centimetre exactly?

Semantically the meaning of money remains undefined. This is the weak point in money and I think it's also the cause of some of the biggest problems economists have to face: interest, inflation and moving big capital from country to country. Poverty is not a real problem, it's part of the game.

The absence of semantics also has a flip side: interests. Do you have a mortgage? Then you pay more in interest than in actual capital payments. Is this fair? Yes, because money doesn't have a definition. It's nothing but a growth pattern, a multiplier. A Euro today is not the same as a Euro tomorrow. If you borrow today, you pay back more tomorrow. If I borrow a loaf of bread, I give one loaf back. The same for a cup of sugar. But, a heap of money for a heap of money, plus an extra bucket of money. Growth pattern!

Let's investigate this a little closer. Imagine, my trousers measure 95 cm. Tomorrow these same pants measure 98 cm. Witchcraft? Have the trousers grown? No. The ruler has shrunk. No economic growth, the currency shrinks. This is called inflation. I call it rubbish. What's the use of a ruler that shrinks daily? Kindling is all it's good for!

It gets a bit more complicated when we move money. You probably think of the "left over currency" in your pocket after coming back from a foreign holiday. This is not what I mean. I mean amounts to the amount of the sum of the monthly income of a small nation. These amounts are moved with the press of a button, electronically, from accounts in Brazil to accounts in Moscow or London. In the old day you'd only move these sums if you moved products to the other side. That's the way it's supposed to be in an economic model, the production chain: Money goes one way as recompense, a payment for goods that come from the other way and change owner. But this Mega transport of funds nowadays does not have a counter weight in goods. There are no consequences, or so they say... But in reality there are consequences, because this "money-fuge" is the origin of a financial crisis. What happens? Investors find out that the money in say Brazil is in danger, because they no longer can pay the interests. So you quickly move the money, before the bank or the state goes bankrupt and I lose my money. When the first pile of money moves, many follow. Whoever is too late and whoever can't get out, sh*t happens. Those that can't get out, can't leave, are, of course the inhabitants. The investors just start over in another country.

So, at this stage in my life I was confronted with a riddle: You have a vision you want to make come true, you possess the rock hard economic truth and you know for a fact that money is nothing more than an undefined unit of measure. That money being a vague unit is the hidden cause of a number of complex economical problems. Ordinary people lost the clue a long time ago. Economists take the problems as fact. Guess twice what needs to be done.

I was trained as designer of information systems. The monetary system is the biggest and most influential system on this Earth, apart maybe from the clock and the calendar. But the design is so faulty that the base, money in fact is an unknown factor. And we trust this unknown factor?

The lack of meaning is a handy tool for banks. When you take out a mortgage, the bank lends you the money and you pay an annual 4% of the amount to the bank in interests.

Question. Where was the money BEFORE the bank used it to buy your house? In a safe? In a bank? The answer is NO, that money didn't exist. When you borrow Euro 100 from the bank, 7% of this money was put in there by a saver and the rest in new money. The money is created the moment you borrow it. To borrow something that didn't exist beforehand you pay thousands of Euros annually! What a joke!

In case you missed this, I'll tell you something about computer programs. Everybody who has seen the program code of another programmer knows how hard it is to change something in that program. If you change something here, you will get an error there, if you change something there, you get errors elsewhere etc. What can you do with such a program? Build it again and improve it. You can no longer repair it.

The monetary system works the same way. If the government introduces an ecotax tomorrow, the day after the whole car and air travel sector gets messed up and a week later the whole economy suffers. Whenever we right one wrong, we get innumerable other errors in its place.

Economics is like an unstable house of cards. It's actually a miracle that it still works, I mean, it's a small wonder you haven't seen through the trick that makes it hobble ahead time after time.

If I had to repair the monetary system as a programmer, I wouldn't try to adapt it. I would make an entirely new design and so write a whole new program. You can't fix the design of the economy because there never was any design. Its functioning is a disaster. Implementing the necessary changes is impossible because that causes the system to collapse. Crash!

Poverty, unjust division of welfare, hyperinflation, fraud, shortage of raw materials, waste mountains, pollution, exploitation, slavery, child labour, waste of energy, climatic changes, commerce wars, drugs, Mafia, homeless people, prostitution and war. In a word, unhappiness. All glitches in the program and for none of these a solution was found. I'm sure you can add some to the list.

Let's see what these glitches look like. Companies are known polluters. And they are allowed to pollute to a certain extent. Losing precious materials costs them money, but this negative balance can be righted by the profits on other products or, on other parts of the same product.

Say, I own a refinery and from time to time I spill some oil. No problem, My Company is right next to a river and nobody notices anything, or at least let on they don't notice anything. I did lose a few tons of raw materials; I lost those in the river. From a financial point of view there's no real problem. The damages are so small that I can make up the loss by the next sales order. Nature doesn't share my point of view however. That colony of swans downstream doesn't like crude oil at all! A fine wouldn't help either with companies that make a $ 10.000.000.000 profit per year. The profit gets a bit smaller, but they will survive. But the swans won't. They can't compensate the damages done to them with the fine paid by the refinery. Here you see that the way the system works does not coincide with reality.

Another example: the greenhouse effect and global warming are getting on the way. More is to be expected. Are we going to do anything about it now? No, conferences stumble along from failure to failure just because the country which has the most arms, the United States of America, pressures the other countries. Otherwise the States would boycott these countries. Or..or else we'll invade Liverpool.

The problems interlock. None of them can be solved because one solution makes other problems even bigger and worse. That's why governments have been sitting around for years, waiting to take action.

*But they, the governments, won't take any action
because they don't dare, because they know it
can't be done.*

That's why around 1995, I decided to re-design the monetary system. But this time in way that the value of the money coincides exactly with the Natural rules and laws. The Dutch Minister for the Environment, Margreeth de Boer, gave me the idea. How did that come about?

Our Organization JP Doe, was one of the few platforms the national organization NJMO had been able to set up. This club met once a year, which now is a part of the National Youth Council, with the minister for the environment. I attended one of these meetings as a depute from Enschede.

Minister Margreeth de Boer talked about "ecologising" the tax-system. Ecotax. She would have liked that very much she said. She had ordered lawyers to investigate the possibilities but the result was, alas, that this tax was not possible: There were too many laws that hindered the implementation.

Before you know it I digress to the fact lawyers made laws that are so complicated that they even obstruct themselves, but let's stick to the subject for now.

The minister promised to look into making the subsidy system greener, now ecotax was out of the question. The other youths of the NMJO thought it was a good ides, but I didn't trust it. I said that that would be too difficult as well. Why don't we look into making the whole money system more ecologically sound? That was a sentence nobody understood. Not even me. But in the years that followed I got to understand what that entails. I had to learn to make better models before I began to fully understand.

1.7 Life cycle

An IT technician often has to solve a determined problem. In order to gain insight in the problem and its solution, you design models. Just like in the world of fashion. A fashion designer cuts the all the sizes before cutting and sewing.

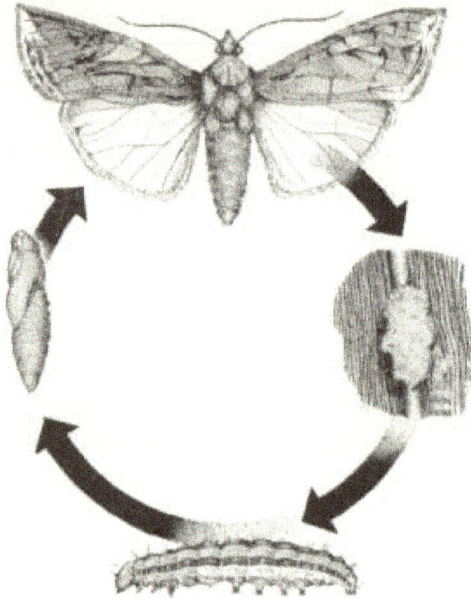

The life cycle of the Moth

I opted to learn more about making models and techniques to make these models. Handy for a dreamer, because than you can exchange dreams with people who can read and understand the same pictures.

There was one teacher who influenced me a lot and his name was Rob van de Weg. What makes his work special is that his starting point is the life cycle of objects. Life cycle is a notion you probably know from your biology classes. It's the same in IT. Objects are born, run the course of their lives and are cleaned up using the "die" function.

The system's garbage collector clears memory so you can let new objects be born in the same space. The same applies in biology. When you throw the mowing of your lawn on the compost heap it is converted into raw materials by bugs and bacteria. This raw material we call compost and looks a lot like earth, the fertile topsoil of our planet.

When I was about top graduate I buried my nose in books about life cycles and drew models about objects in a hospital. In the back of my head the mission to re-design the monetary system, the mission I set myself when I met the Minister, was waiting. But isn't recycling part of that economic system somehow? How did they think about that, back then when the king minted his first coin?

1.8 Production Chains

Instead of defining money economists give a description of the economic flow, the production chain. Goods are produced and move from manufacturer to manufacturer, until the end product goes to the consumer. The money goes the other way, from the consumer upstream as it were to the producer until it reaches the supplier of raw material. Every sale implies a money-flow to the seller. A big part of this goes to the seller's supplier, but every enterprise keeps a part of that money, the margin.

Here you find three production chains in a picture. Look where the end product, meat and vegetable, end up. It doesn't say. After retail it goes to the consumer, OK. And after that? Right, literally flushed down the toilet, sewers and incineration. And after that? Dumped in the sea, or burned to an unknown material. Also nothing is known about where the tin ends up.

This model, the chain of production, is the basis of the economy. It has two shortcomings. The mining part exhausts natural resources. On the other side of the column there is the waste mountain. It consists of the same material as the stating material, in useless, used up form.

In the last phase of this cycle the substances are dumped into the sea and air. But where is the money in that last part? Nowhere. The great Unknown. It was never thought about.

I would call the value of the finished products negative. It costs money to use the sewage system. It costs money to collect domestic waste. Everything costs money, you'll think. Logically, because garbage has no value, it's value-less. I think that's strange, because the waste of your toilet can be recycled. It should be recycled. That waste supplies precious raw materials for our horticulture. No, the value of our excrements is zero. You pay for the trouble, i.e. energy, of the "clean-up".

Chains of Production

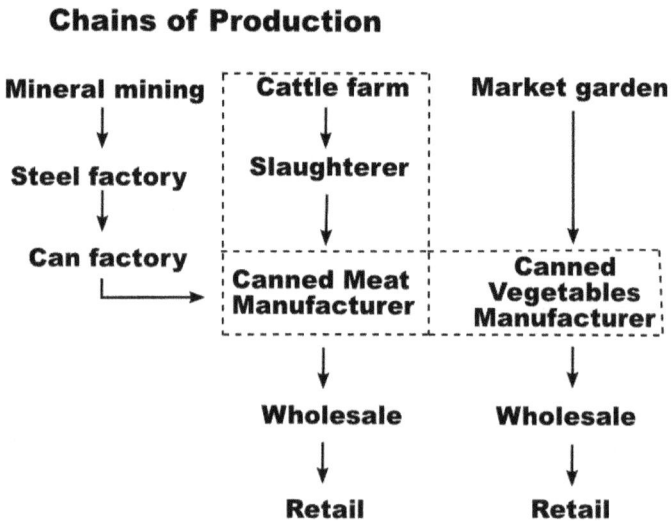

Source: Onderneming en omgeving, SMD Leiden 1998

Three chains of production

The shortage in raw materials and the waste
surplus are caused
by this production chain, because recycling hasn't
been taken into account
in the design.

You will say now: Right, but recycling is being done these days? I collect my old newspapers and they are used again? Sure, but that is not thanks to the chain of production. This is a case of two connected chains of production: the regular industry and the recycling industry.

That's why every product needs its own recycling system. The (in)famous PET bottles were recycled at the supermarkets up to 2006. That obligation ceased to be in the Netherlands. Paper recycling is organized in a different way in each municipality. Often it's collected on a fixed day during the week, through organizations such as the local scout group or local volleyball club. You don't get any money for your old papers. Only when you hand in 1000 kilos at once it's worth your while. Tin cans are retracted from the waste with big magnets. The most important, organic waste is collected in special containers and composted. But our faeces... nobody ever mentions them. Flush them down and don't look back is the motto.

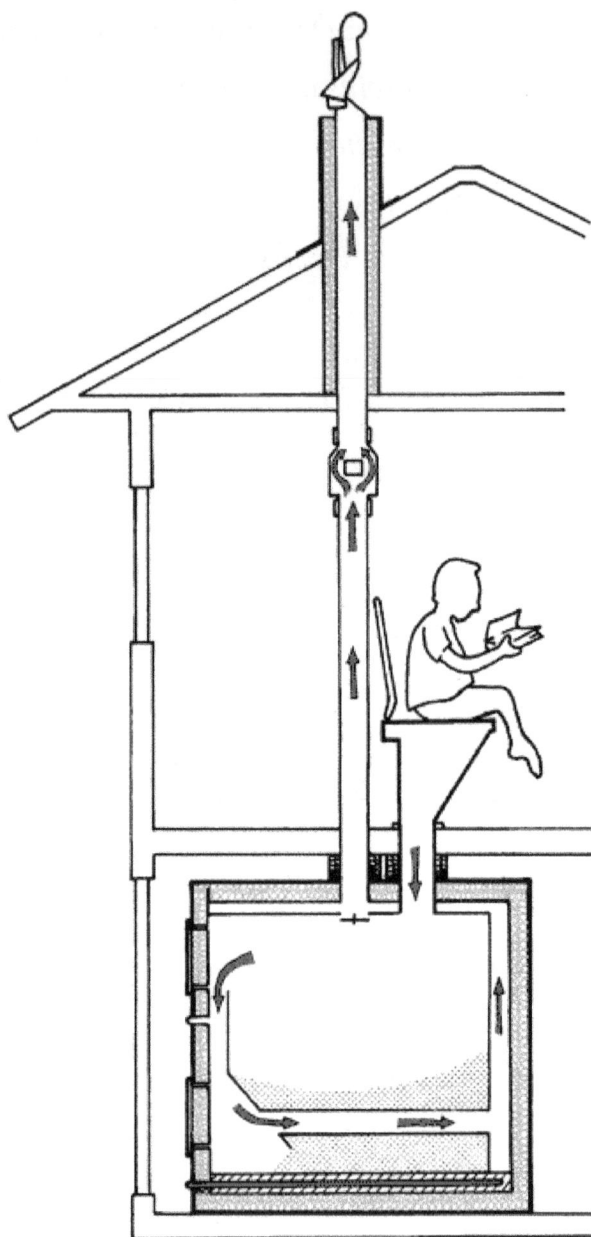

Owners of a compost toilet appreciate humanure

1.9 Sun and Earth

In a nutshell my life was as follows: I'm 23, live in a world that is dominated by powers and the biggest power is money. The world itself is the main victim. Even though all world leaders know we are up to our neck in trouble, they can't do anything about it because the whole system would collapse if they did. Since the group that wrote the Enschede Vision fell apart, it seems I am on my own to bring this planet back in balance.

My girlfriend at the time presented me with a nice necklace. It had the sun, earth and moon as a trinity on it. The sun heats the earth with her warm beams, the earth is life itself and the moon is there and influences the tides, the weather and human rhythms.

Just like in the Enschede Vision I saw in that pendant the difference between energy and material, everything contains both, but you can look at everything from two points of view. The "material" point of view and the "energy" point of view.

In high school they make you solve problems where flowerpots fall of a roof at a determined height. How long will it take for the pot to hit the ground, and at what speed? You can solve this in 2 ways. Using formulas with space (distance, meters) or formulas which use energy (height and speed). Both calculations obviously will give you the same results.

Albert Einstein added to this. Everybody knows the formula

$$E = mc^2$$

But not many people know what it actually means. It's about exactly the same difference between matter and energy. E is the amount of energy. M is mass, usually called "weight" and expressed in (kilo) grams. The c stands for the speed of light, an enormous number. The 2 is " square". Or $c^2 = c \times c$.

What this formula basically says is that you can transform energy into matter and the other way round. You can calculate how much energy is involved if you know the difference in weight before and after the process. The thing is that these processes take place in the sun or in nuclear reactors, small suns on Earth. In brief:

There's a huge division between

matter and energy.

This division is so great that it takes a sun or a nuclear reactor to eliminate it and transform something from one into the other.

In principle you always keep energy, and also matter, no matter what processes take place. Only in this special case transformation will take place, but that is an "un-earthly" process.

This division between energy and matter, symbolized by Sun and Earth seemed an excellent starting point to base the new economic system.

Nature's laws are universal and, as opposed to lawyers' laws, they can be described in exact terms. If we start from these natural laws, we automatically get a system that makes sense, instead of a money system that is based on an illusion.

The Moon

The pendant had three elements, the Sun, the Moon and the Earth. Once I understood Sun and Earth, it took me at least six months to fit in that little moon. I had to revert to "life cycle" to do that.

When you look at the garbage can or and the organic waste and paper container that are probably next to it, you see one big difference, one container contains stuff that gets to be recycled, the contents of the other box we'll burn later on. What happens to the fumes and gases that are released in this burning process is unknown. They end up in the air, the water, on the land. They aren't recycled as such.

It would be nice if you were rewarded when you recycle and punished if you don't. This is already happening to a certain extent, but due to money's own special mechanisms you can make up for loss (negative balance or a fine for pollution) by a big profit on the other side.

You can solve this problem using a separate symbol. The symbol "Earth" represents the amount (kg) of living, recyclable material (organic) and the rest, the non-recyclable waste, is represented by the moon.

So now we have a 3-symbol system and I called it Econism:

A system where the price is always

in accordance to the value

according to nature

Eureka. Not even 25 yet and I had discovered Columbus's Egg!

1.10 Who wants the egg??

I spent 3 year on thinking how to solve Earth's problems, and once I finally had the solution, nothing happens.

It seems that no-one was waiting for this. Who wants something to do wit Sun, Earth and Moon? It's all way to vague. I admit, I pushed this story too hard to people who weren't ready for it. They told me to write a book. Right, writing a book... Surely I don't have time to do that!

So I ended up back at the NJMO. Meanwhile the management of the organization had changed, and there was hardly anybody left who knew who I was. Of course they wanted to know what I wanted, but the answer came:

Oh Dear, we currently don't

have a project in that..

I could have known back in 1993. Nothing has changed in 2006. People and their organizations follow their own goals and projects. The NMJO did projects for which they received grants. Those grants took a lot of work and trouble to get. Money came in or it didn't, you never really knew. If the money came, the project could go on. And then you show up with something like "ekonism"... No thanks. We don't have time right now...

Other reactions were along the lines of "go to this and that organization, because they're doing something similar". Once I got there I wasn't understood or they were too busy. Often I was dealing with people who weren't the original visionaries of the club. They sent me to their "leader", but I'd stopped returning phone calls by then. I didn't want to be sent from pillar to post, and I still don't.

1.11 Matrix

Later the film "The Matrix" came to the cinemas. Like many people who recognize their own struggle in that film, I saw that people let computers decide their fate. We let our lives being led by automatons that are steered with numbers, the economy. Computers do the work, but we filled them with numbers, money... My bank account, my salary and my rent. I work to keep my bank account to certain level so I can keep paying the rent.

Capitalism is an illusion, a human invention that turned us into slaves. The first part of the movie is about breaking loose from the influence of the matrix, capitalism.

As it turns out, many people aren't ready to be free. They believe in this fantasy world and can't do without it. There are more and more individuals who are ready however. They are made "half free" and are offered the choice, the red pill or the blue pill. Going back to the matrix or stay out of it and see just how deep the rabbit's liar is.

"Reloaded" is the title of the second part. Re-loaded. Here you see the struggle of free people against the system continues, but in order to fight the fight they are sent from pillar to post by players inside the system. Neo listens to the advice given by "The Prophet", a character out of the Matrix. She sends him to the Merovingian, another character in the Matrix and on to The Key Maker. The architect, ditto. When the whole free city of Zion is butchered by machines, Neo tries to save his girlfriend, who earlier in the story tried to save him. Love seems the strongest factor.

The principle of "Reloaded" is typical for the environmental movement in Holland. All nicely done, but to get things done we need money and so we go to the government to get money. The same government we should be suing. You don't believe for one second that the Environmental Federation "Zuid-Holland" (a union of environmentalists and activists) is in favour of yet another motorway "under certain conditions"? How is this possible? They've been reloaded. They are fighting against the Capitalist System but have become dependent of that same system. Their staff also have families, a mortgage and a lease-car. Reloaded.

The third film shows how the battle can be ended. Unification and love. Clear and simple, Love is All.

1.12 Towards Politics

If nobody wants to hear, I'll go and bring it myself. In the elections there was a party which had my name. Natural Law Party. I didn't know anything about them, but voted for them nonetheless, just because of the name. The name Nature Law and Ekonism matched exactly!

Obviously the party didn't win any seats in the parliament. A couple of phone calls later I was invited round and I was offered a place in the national management of the party. What do you mean, "You need people?

As it turned out the Party was a group of people who meditate and use "consciousness technology", The Maharishi from the town of Vlodrop in Limburg seemed an example to them. The leader of the group was his right hand. They had fantastic food and were really nice people. They wanted to hear my story, but then they were rather stuck with it. There was a subtle reference to it in the next election program, but then things started to go askew. Power struggle, War, fighting, Big mouthing... What on earth could cause that?

1.13 Crocodile babies

Of course I meet lots of people, mostly men, who can't or won't believe that a world the way I see it is possible. No, that's not how the world works. OK, I know you have to keep your feet on the ground, but on the other hand, the way I see the world influences the way I live in it. If you believe something can't be done, it can't be done. In your world at least. And that is the only world where we meet... so far.

Often I hear this complaint:

I'm all for idealism, but when push comes to shove I still have to choose between me being hungry or the other fellow hungry. Everybody chooses food and let's the other guy go hungry.

You recognize this thought pattern?

A biologist once explained me that the human brain have a kind of primitive brain that is only used for the most basic task: Survival. This brain is a leftover from the time when human beings, at least according to Darwin, were reptiles, and that is why it's still called reptile brain.

If you can't think beyond the point of "eat yourself or go hungry", you only use your reptile brain. I'm not amazed that this is exactly what you see in a crocodile pond: Big bites and big mouths. No leftovers for the next guy, the whole world is my enemy! Bite!

Of course it's up to the individual whether he or she chooses to live with people who have this dog eat dog mentality. Do they really not care about the others?

That depends on who "the others" are. Even the cruel, selfish and hard crocodiles have a weak spot. Mothers. They have that special gift to protect their young for the first year and a half in their safe mother-mouths.

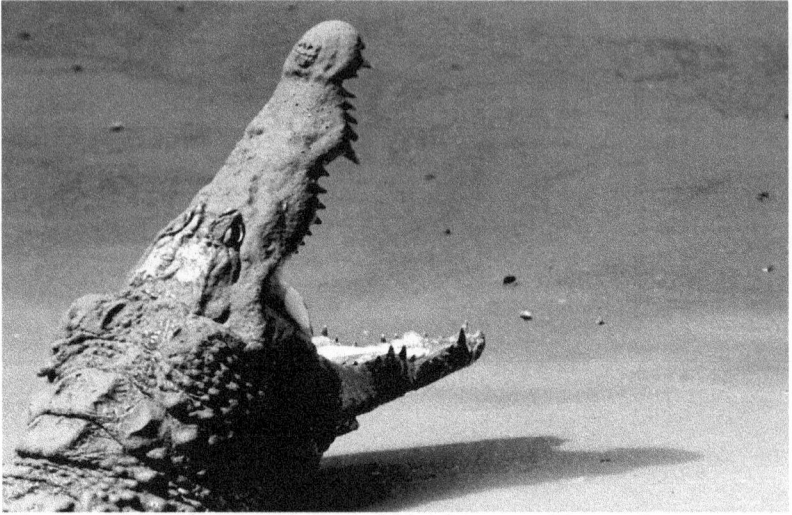

Figure 3: It's a dog-eat-dog world in the crocodile pond. Photo by Kevinzim.

Apparently the crocodile is a cruel monster that eats everything and everybody but herself and her young. If the whole world is their enemy, as I said earlier on, the young is not part of "the world", but part of the "mother-self"! It's me-and-my-young against the rest of the world. And especially, the rest of the world against me and my young. Because that's the way it works in the crocodile pond.

It's the same with human beings. A mother shares her food with her young. You could say, she is her child, she's one with her child.

Do you recognize that? That feeling of unity? Do you feel it with your children, your pet, your partner? Maybe the feeling of empathy? To be able to put yourself in some one else's position? These are functions of the limbic system. This part of the brain is a lot more modern than the part we have in common with the crocodile. You could say that this is what makes us human. We are higher animals, not crocodiles that kill each other to get the last bite.

The most important is the border between "me" and "the rest". The crocodile only considers itself as me and separates thus from the rest of the world. The crocodile mother also reckons her child to her "me" and sees the rest of the world as hostile. Nowadays there are more and more people who feel "one with the world". We all breath the same air...

Lots of people see me as some kind of "eco-knight", a "green Don Quichote", because I don't own a car and I am vegetarian. I don't want a car because they smell. When you drive yourself you often don't smell your own car, so I didn't take that decision purely out of self interest. No, the world around me and I are one, and I don't want to pollute my "self". It hurts me to see the state of the world now. I feel the pain of the Earth. And that is why I say there is no border between me and the rest of the world. Maybe this example is a bit far-fetched, so let me find another, easier example.

When I was a child we saw these spots on TV at Christmas. Hungry African children with swollen bellies. My mother used to say we should be thankful that we have it so well. Now I know that our abundance is their shortage. And I am not thankful or happy for that. How can I be happy with our "welfare? I can't any more. I am unhappy with our "abundance", we should share it with those who have shortages.

Again this is an example of me and the other. There is no difference between me and Africa. I don't see the difference. Who wants to see the difference at all cost, is deluded. I breathe the same air, drink the same water and live on the same planet as any other being. I am in the mouth of the world-mother and my mother is a crocodile.

1.14 Escape from the crocodile pond

Why is our economic system exactly like a crocodile pond? Why do we work based on struggle and competition? Why do companies take the last bite and let other companies bleed to death?

They do this because the economic system is connected to our reptile consciousness. They use the most primitive functions in a body: the own survival. Either you eat or I eat. And because we're all crocodiles, this is how we treat each other, that's the norm. It's been like this since the dawn of time – History of the Western World often doesn't go back any further than the period in which a region was conquered. Yet there is an exception to this rule. These last few years an industry sprouted up that is based on the concept of sharing. No fighting but co-operating. For this we go back to the story of my life.

About the same time I joined the Natural Law Party, I "discovered" the computer. Making models was fun, but I wasn't really drawn to computers. Only later, when I was working for Internet providers I came across free software. This is an important stream. Against the tyranny of Microsoft, a world community of already more than 1000 000 programmers succeeded in writing a complete system with all possible programs imaginable.

This started already in the eighties, when Richard Stallman wrote the first copyleft license for free software, the GNU Public License (GPL). Later he wrote a number of programs to write programs and published them under GPL. That allowed everybody to make their own free software. In the early '90s Linus Torvalds added the 'Linux' kernel. One year later this kernel was brought out under GPL license and there was a usable computer system which operated with free software. Soon this system was implemented by internet providers and research facilities.

The contrast between this and the commercial world of software is huge. Still today thousands of "specialists" are trained and they only know Microsoft Windows. If you want a program, you have to buy it or download it "illegally". Most of the time you need a more or less expensive license. You don't get a copy of the source code. You don't have the right to change even one character in the program. You can't divulge the program, can't share it with your friends... Of course information always finds its way, so people share programs anyway. But form a legal point of view the software isn't free.

Stallman says free software has to guarantee 4 rights:

1. the right to use the program

2. the right to study and spread the source code

3. the right to distribute the program

4. the right to distribute and make public your changes

This way you have full control over your computer. By exchanging the code a community came into being, via the Internet. A community where people work together to reach a common goal that seemed impossible. Right now you could run your computer with free software and not miss a thing. Even better, you will have a lot more!

Microsoft has bought every company that was leader in their field. Only the Free software world escaped its claws, because it's not for sale. According to some people the value of the kernel is about 2 billion Euros. I don't believe this because money doesn't come into play here.

Besides the release of free software there is a new trend: the free word. Traditionally books are covered by copyright: the writer (or rather the publisher) reserves all the rights of the publication. Nowadays there are books under copyleft, one copy for whoever wants it. Wikipedia, the Internet encyclopedia, is an example. Nature magazine states that the quality of the articles published on Wikipedia are similar to the entries in the Encyclopedia Britannica.

Of course, the book you are reading is also available under copyleft license from CreativeCommons.org. Go to the nearest photocopier and copy me!! To change anything in the text you only need the Openoffice.org word processor. And also this is adaptable to your needs.

The worldwide programmer society showed me how to deal with power. Create your own rules. It can be done. You are not obliged to surrender the rights to something "above" yourself, like a company or a boss. When technicians don't agree, they reach a solution without the need for a judge or referee. Working together and sometimes working by yourself. They don't surrender the control over themselves, the power. Everybody "owns" their own work. How great the contrast with our "modern" democracy?

1.15 The illusion of the majority

The Dutch - and probably most democracies- let themselves be governed by the minority of a minority. One third of the people don't even bother to vote even more, because they have lost all contact and trust. In fact the seats belonging to those non voters should be left unoccupied, because now a minority takes all the power.

Two thirds of the people determine who gets a seat in Parliament. But, people vote for Party candidates. The Dutch constitution doesn't say anything about parties, yet the political field is mainly made up out of parties

Say, a number of parties divides 2/3 of the votes. 1/3 of the voters didn't show up. 3 big parties holding 55% of the votes make up the government. Conclusion: the government is backed up by only 2/3 x 55% = 37% of the electorate.

In important matters the majority in the Parliament has an important role to play. But within the parties Fraction discipline is the rule, MP's are expected to vote the same way the majority of their fraction votes. Even if you are against. The constitution (article 89 of the Dutch constitution) states that Members have to be able to function without counselling or burden. Apparently this isn't important.

Say that in a given matter 60% is pro and 40% against. The decision is supported by 60% x 37% = 22% of the voters. Yes. Are we to build a new power plant? 22% decides. Does Minister X have to resign his post? Again, 22% decides And you thought democracy was about the will of the majority?

Those 22% is a rather optimistic calculation. A lot of people aren't allowed to vote on grounds of their age. They aren't "mature" enough to vote. People until the age of 18. As it happens these are the same children that can work MSN, unlike our ministers, even if they have ICT in their portfolio.

So called foreigners do not participate in national elections. They were born on the other side of an imaginary line we call border. After this line "the exterior" begins, a region where a different power structure rules. You don't have anything like it on the Internet. There is no "abroad". You don't even notice whether my website is in your own country or not.

Have you ever seen a border? Never, I'm dead sure. I've seen border markers, or fences. The pavement looks different in Belgium than the way it looks in Holland. I saw the Wall of Berlin. That wall was placed a long distance before the border. But there is no border as such. The concept only exists in your brain. Forget it, there are no borders. Abroad is in your head and foreigners you only see on TV.

Anyway, the government is only backed by minorities. But, is that the whole story? No, because the division of power within the parties is at least just as strange. Even though the parties are responsible to govern every body, only the so-called members who carry a certain weight determine the order in which the names appear on the ballots.

The selection of the members of the government is done through a completely non-transparent process of meetings. The members of government are appointed from a list that wasn't available beforehand. When the list is made up, it gets approved by the Queen and the Parliament. Also that other important document, like the all-important " Ruling agreement", is made up in back rooms, inaccessible for press and citizens alike. We would like to know on which conditions these negotiations take place.

I'm sure the Queen is a nice woman, but I don't recognize the State of Holland and the money with her picture on it, the Euro, as the only one.

I claim our freedom, our birthright to a free world and a free existence, without obligatory slavery to that dictatorship called "economy".

I keep going, every day, step by step towards what eventually will be unavoidable. And when I let go, I haven't forgotten anything.

1.16 Duality

When you take a daily interest in the news, you surely are familiar with the notion of "Duality", a description of the whole by emphasizing the differences between the parts that make up the sum. Politician A fights Politician B. Country A has a beef against leader B. But C was such a bad dictator and had to be disposed of, even though we didn't find any Weapons of Mass Destruction. Struggle, War, Fear, Terror.

Sadly the news in the mass media is nearly 100% made up of this kind of "news". The best thing to do is to block it out. Neither party is right. In case of war or Warrior speak, none of the parties has the right to be "right". Who accepts the truth, embraces the opponent and forgives him.

It's time to find a way out of this duality. As far as I can tell it can be done with a Trinity.

Part 2. Colourcash

2.1 Introduction

An important lesson is "letting go", this means being able to literally let go and forget about it. I kept the whole "ekonism" in a box and on an old website for a few years. Until I told Suzanne everything about it, and as it turned out, there was somebody who wanted to get a grasp of the idea. We changed the concept a bit, because people don't really want yet another "-ism", and rightfully so.

The symbolism is a bit too vague as well, it could be a bit more concrete. One Sunday morning we adapted the whole thing and "colourcash" saw the light of day. To fully understand this I will first explain how colours work.

Everybody's heard about the primary colours, red yellow and blue. With these three colours you can make all the other colours. Paint filters light, the more colours paint you throw together the darker, browner it gets. When you use coloured light instead of paint, you use another set of primary colours.

(Computer)screens, televisions, cameras and projectors use red, green and blue. Also the colours in a website are these same three colours. Using these three colours you can make all colours the human eye can see.

Black-and-white is a lot easier. When you draw a picture and save it in the gray-mode, you only store the amount of white in each point. The more, the whiter. No white, all black, max. white, all white. All grey tints are a balance between those extremes.

In the RGB-colour system you use three values, a single one for each of the colours red, green and blue. In a black and white system there's only one single grey-value. This reminds me of the way money works.

Imagine, you live in a world that places value on "height". The higher something is, the more value it possesses. It can be beautiful or ugly, edible or poisonous, round or square and high or low. Energy-rich or energy-poor. Alive or dead. Recyclable or non recyclable.

In the example the height is the only factor that determines value. The rest doesn't matter. A keyboard is rather low, about 5cm. Just a bit cheaper than a tennis ball. A racket falls down and lies flat and so is cheaper than a keyboard. A pair of golden earrings is even cheaper, because they are lower.

Do you understand this vision? All values are flattened to a single value meter. This value, height is not really important to

 1 dimension: a line

 2 dimensions: a plane

 3 dimensions: a cube

Figure 4: Dimensions in a spacial world.

59

us, but who are we to judge someone else's values?

Our own value system, money, is just as primitive. All values are flattened to a single value, one dimension, it's money-value. This we call black and white money; of all the values that we present originally, only one is fixed and kept. Just like black and white photography and B&W TV.

We would get a lot further if, just like with colours, we had three separate values: one for red, one for green and one for blue. Then we have three dimensions, three completely different things that are not to be mixed up. If you want to know how big a cube is, you don't add height, width and length!

We will come to the meaning of the colours later. For now, when you use this system, you find out that "more" isn't "better. We are looking for balance, equilibrium. This is easier to find if you use less energy (red) and recycle everything (green). That's easily done if you co-operate. Instead of the classical commercial production chain we use the economic cycle so recycling is incorporated right from the start. This way the whole economical game shifts. Ale rules are re-written. This we call colourcash.

2.2 Colourcash

2.2.1 Introduction

What follows is a philosophical summary of the most important point that make up the base of colourcash. Colourcash is not just a coloured sort of money. It's a holistic approach to living. You can't take out "the juicy bits" and leave in the hard stuff... It's take-it-or-leave-it package.

This package isn't really open to discussion. It's like the seed of a tree... Small, yet big enough to contain all the necessary elements to grow. You have a say in where to put the tree, you can choose the earth you plant it in, how to prune it, these are the topics we treat in the third section, like tax and insurance.

Colourcash is a system
where the price of
everything is
represented with the
colours red, green and
blue.

We can differ in opinion and that's OK. This way you can choose universal colourcash, yet give it your own "meaning".

2.2.2 Everything lives, everything is nature

Everything alive has a beginning and an end.
Everything that goes on living, lives in a cycle.
Because of the cycles, generations can succeed one
another.

This is true for man and animals, but also for plants. It's even true for minerals, planets and stars. Their life cycles are longer than ours and these life forms are often considered as non-living. They are not built of the same elements we are, carbon, water and oxygen. But so what? These elements are nothing more than a vehicle for energy.

Because we know they have a beginning and an end, they are alive.

2.2.3 The Truth

There is only One Truth, but it can be known
in very many forms.

We, as human beings only need to investigate this truth. It's not about who has the "absolute truth", Nobody has. Everybody is subordinate to the truth and the story, the point of view of each and every human being contributes to this truth to the same amount.

There's an old Dutch saying we use when two people are fighting about "who is right" ... the truth will be somewhere in the middle. This is only a part of the truth, the truth is a duality of both points of view. Both are right, from their point of view.

2.2.4 The unit

A unit is the smallest part that can't be divided
without losing the whole.

The potato, man and bicycle are units. Also colourcash is a unit. You can't just take out parts because they suit you.

The whole can be divided using a duality or a trinity. We are familiar with duality, right/ wrong, good/bad, left/right, more/less and black and white. The Yin-Yang symbol teaches us that dual partners can't live without their counterparts.

Figure 5: Yin-Yang, a symbol of duality.

Colourcash prefers a trinity. A trinity keeps the unit united.

2.2.5 Trinity

All trinities are the same

It doesn't matter whether you chose the trinity Red-Green-Blue or another trinity. You are recommended to think for yourself first: What could be the meaning of the colour Red? Obvious, isn't it?

Red: Energy.

*Light,   Life,   Love,
Warmth, work, Plants,
animals, seeds, people and
arts.*

Red grows during life and diminishes towards the end, just like the sun rises, turns at its highest point and sets again. Red is the heath of the stove, the light of a lamp, the force of a shot and the love of your wife.

*Green: living   matter.
Everything   that   gets
recycled.*

You find green in food, water, earth, compost. Green is the building material of life, the Earth's womb, your cradle and your grave. Green is the leaf on a tree, even if it's yellow the day it falls. Green is the pure rain water. Green is the manure you spread on the land.

Blue: all matter that's been taken out of its life cycle and isn't deliberately recycled.

Usually substances are released, or dumped, in the air or in the water. Sometimes these substances are imported from far away so it doesn't automatically return in the same cycle. Blue is the colour of water, pumped up from deep beneath the Earth's surface, blue is also the colour of the garbage bag , taken away to be incinerated. Blue is the colour of exhausts, because we burn a lot more oil than the planet makes new oil.

Now you know the meaning of the colours and I'm sure you will have noticed that the colours have been based on the difference between energy and matter on one hand and matter on the other and the division between closed and open cycles. But when we want to calculate, we need a base. After all, how long is a meter?

2.2.6 Definition of the unit based on sunflower oil

The energy in 100 gram sunflower oil we consider 1 Red. That is 3400 kJ or 810 kcal. The substance we reckon as 1 Green. When you change the oil chemically into non-biodegradable plastic, or you render it unfit for recycling in any other way, the substance is reckoned as 1 Blue.

This yields the following definition:

- **1 Red** is the energy in 100 gram of sunflower oil, i.e. 3400 kJ or 810 kcal.

- **1 Green** is 100 gram of substance within a closed life cycle.

- **1 Blue** is 100 gram of substance not within a deliberately closed life cycle.

2.2.7 Alternative definition for the US

The United States and possibly other countries have not adapted to the metric system with its kilograms and meters, but use imperial or United States Customary system. Using 100 grams of sun flower oil may be a bit unclear. A redefinition in pounds is as follows:

- **1 Red** is the energy in 0,22075 pounds of sunflower oil, i.e. 3400 kJ or 810 kcal.

- **1 Green** is 0,2207 pounds of substance within a closed life cycle.

- **1 Blue** is 0,2207 pounds of substance not within a deliberately closed life cycle.

2.2.8 Balance

Looking for long and short time equilibrium and balance is the goal, the road to success.

You can recognize balance by the long-term, dynamic equilibrium between Red, Green and Blue. Every year the balance is made up and debts are reset to 0. But, all transactions and the yearly totals will be stored and publicly available. It doesn't make any sense to accumulate debts. The unbalance became clear a long time ago. We only need to know and acknowledge the unbalance, not to rectify it.

When you use the ordinary Gregorian calendar, you will probably choose 31 December or 30 June. If you use a 13 moon calendar, keep July 25 as a date, it's a holiday by the name of "day outside of time". [Toonen].

2.2.9 Money in the production chain.

Let's go back to the base of our present economical system. The production chain. In the picture you see that products are traded from raw materials to consumer. The products travel one way, while money goes the opposite direction (up stream). In this figure, products go up and money goes down in the production chain.

The picture uses pie charts which show which portion of the "consumer price" is being paid. Obviously the consumer pays the most, the shopkeeper keeps a part, de manufacturer also keeps a part and the rest goes to the suppliers of Raw materials. This model is obviously a simplified representation because in real life there isn't a product which has such a simple production column. But it works as a model.

2.2.10 Colours in the economic cycle.

Now we have seen how money moves in an economic cycle, we also want to know how Red, Green and Blue money moves. The notion of a circle is just as elementary as the notion that Earth revolves around the Sun and not the other way round. It might take a while before you see the picture. On the other hand, once you got it, you don't need this book any more.

Reality gives a clearer insight than a theoretical model.

In Figure 6 we see the same players in the production chain, from the supplier of raw materials to the producer to the retailer to end consumer. The destructor or reducer is a new player. The destructor destroys the waste-stream that comes from the consumer and converts it back to original Raw Material. This can be collection of used glass, paper or tins. It can be collection and composting of organic waste. In any case the destructor closes the cycle.

Thanks to the reducer or destructor we can close the cycle. The result is that raw materials and the material in the product are valued as "green". We don't need Blue yet. In this cycle, the value of green is 100%, because all materials are recycled.

The easiest way to accomplish this is if a producer decides to become the main reducer as well. This situation is often used in the Cradle to Cradle business paradigm, a way to bring nature into business. For now, we will continue to bring economics into nature. The Cradle to Cradle concept is further explained in the last part of this book.

The value of "Red" varies during the life cycle just like the value in Euro varies in a traditional production chain. Raw materials don't have any or very little energy. The manufacturer adds a lot to this value and the consumer pays the full price in Red.

In Figure 6 products travel clock wise while RGB coins travel counter clock.

The pie charts are used here to give the idea, as is usual in models. The real price will be determined in real life. In principle the "Red" value at the beginning and end of the cycle is "nil" or very low. Just like the Sun rises , gets to its peak height at noon, and sets at the end of the day.

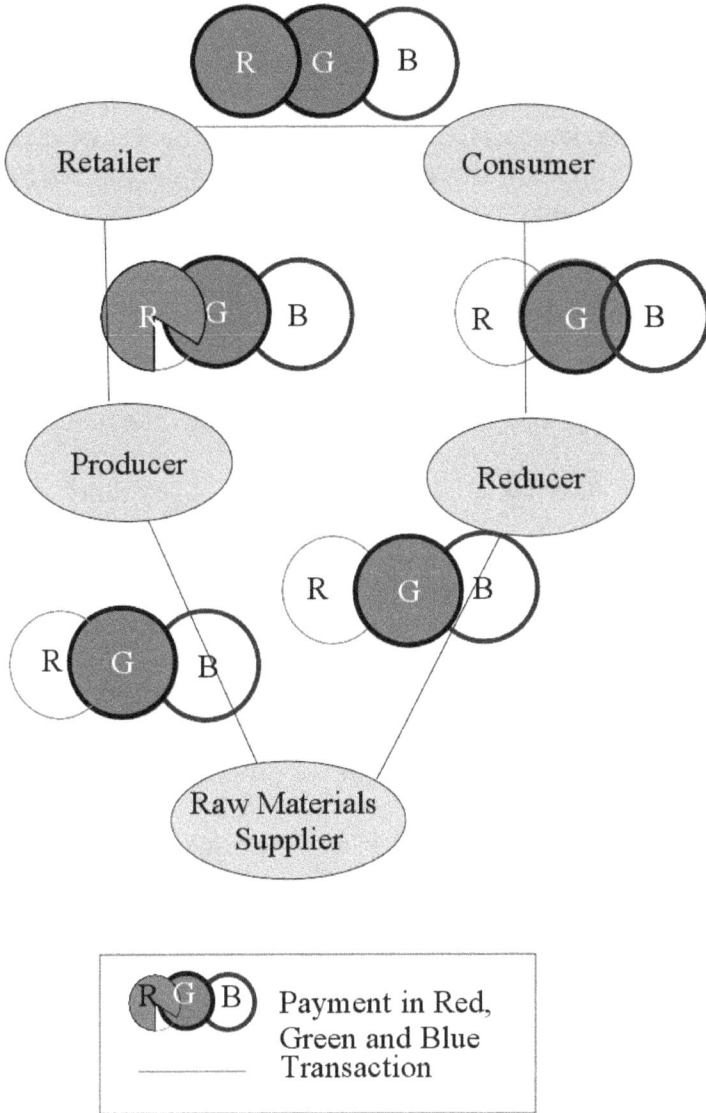

Figure 6: RGB Values in a closed life cycle

2.2.11 How does Blue get in?

It gets a bit more complicated when a product or part of a product no longer gets recycled. Then Blue enters the picture. Blue takes over part of the role of Green. Green is used for matter (read weight) that's inside the cycle. Blue is what you pay when you know this matter leaves the cycle. In both cases you measure the weight. 100 grams is either 1 Green or 1 Blue. Someone always carries the brunt, some one ends up being responsible for a product getting out of the cycle. Someone who's stuck with Blue and has a shortage of Green.

In the following example we pretend the consumer is responsible for half the non-recycled product, the other half is recycled. Apparently the consumer made the choice to use the product in such a way it couldn't be recycled any more.

We see an extra player coming into play , the Blue Pile. This is a virtual participant. The Blue Pile represents all garbage dumps, incinerator and other place where waste is dumped into the environment. Whenever it's outside the cycle, it becomes Blue. In this case the reducer bought the waste product of the consumer for R0 G 1/2 B 1/2. The consumer paid R 1 G 1 B 0 when he bought the product. This means that the consumer is stuck with an end balance R-1 G-1/2 B1/2. The deficit in Red can be made up for through labour, as usual, but the loss of Green can't be recovered.

That is exactly how things work in the present reality, when a product gets out of the cycle, it can't be compensated for either. The off balance stays in the books, until the waste is cleared or an equal amount of matter, with the same weight, is re-introduced into the cycle form outside the cycle.

The combination of the use of these colours and the way they behave in the cycle gives you a clear idea of reality when it comes to energy and the recycling ratio. When we compare to the production chain again we see there is no connection between price and cycles or energy. In fact, money "promotes" the destruction of raw materials and energy. The model shown here is a, in a "model" sense" almost perfect solution for the "internalizing" of environmental cost. But to be able to use it we need to respect some rules of conduct, some basic "life-laws" which we will look into before we go to examples with numbers.

2.2.12 Respect the Cycle

People organize themselves in cycles which have "space" for everybody. Nobody has the exclusive rights to the middle of the circle. All production processes are organized in a cycle, otherwise they can't make up an organism that produces an organic end product. This means that recycling and de-construction is taken into account at production.

Neglecting to recycle is not "a capital crime" but is merely indicated with the colour Blue instead of Green.

Watch out for Smurfs!

If you only get products that are "Blue", you run the risk of collecting so much Blue that you change, as a manner of speaking, into a Smurf.

When your property has run out of fertility, you can't cultivate any vegetables, you are a Smurf. Your animals won't be fed and your children will go hungry. You will depend on your neighbours They will help you, unless they have become Smurfs themselves.

2.2.13 Responsibility and transfer

There is no possession in colourcash. The Earth is the Earth

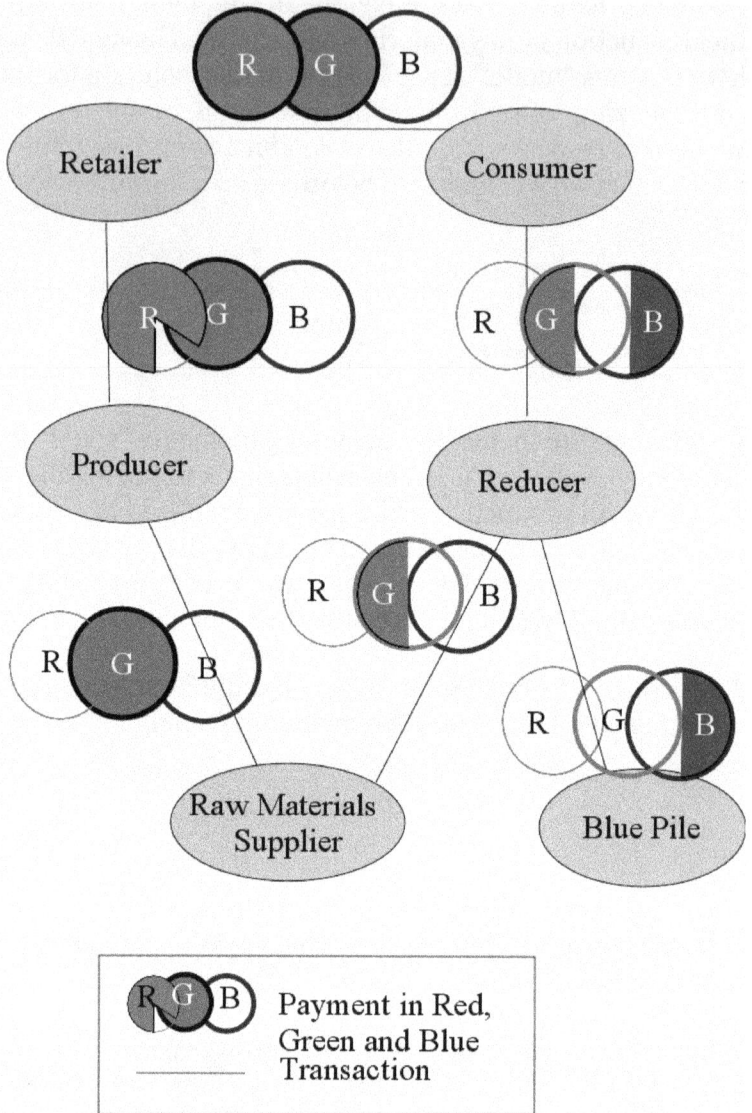

Figure 7: RGB Values in a partly closed life cycle

and all its inhabitants can equally claim to use it. Everybody born on this planet is "earthling" and has the same rights. Every Earthling can get the responsibility for certain goods, fields, animals. But this is not "property", because there is no monopoly.

The responsibility can be transferred in a "trade transaction". The place of goods can change, but there is no notion of possession.

Compare it to the situation with your children, they're your children, true enough, but you don't "possess" them. You can't do anything you want with them because they are independent, worthy beings.

2.2.14 Openness and transparency

The complete truth can only be known when there is complete transparency. The common situation and the individual situation can only be known when all data about all transactions are public access.

2.2.15 Time, space and money

Labour can't be measured in time, only in space. For example in the case of transport speed isn't the issue but rather how much labour, or force, effort, is needed: Fast or slow both require the same amount of power. That is a physical law. The faster the transport, the more energy is needed, often leading to waste. Slower is more economical, however, we want or tomatoes to get there fresh. Deciding the ideal mode of transport is the trick.

Time is Art

A Maya-wisdom that could also mean: being on time is the art. This also means in any case that time isn't money.

Space is money

The Sun irradiates the entire Earth and it's up to this Earth to catch and store this energy. This is done in the form of Life. The more space something takes up, the more energy it captures, the more Red it turns.

So it's essential to make optimal use of the Earth's surface. Optimal means in our case: Energy-producing. Plants, Life. Fields produce Red, food. Forests are life, so Red. Useful. Distribution centres, roads and offices also take up space, but do not produce energy. This is not affordable. The surface built up with lifeless objects shall diminish to smaller, more reasonable, balanced proportions.

2.2.16 Patents and discoveries

All information is available for free. Even "inventions" are nothing more than a discovery of an already existing universal law and don't belong to the person who discovers it first. Patents and discoveries are not recognized as an individual right, but rather as a common responsibility.

2.2.17 Law and politics

All rules have to be accepted by the circles, Bigger circles have the option to adopt the same rules. Rules are basically decided from the bottom upwards. Everybody agrees in principle to those rules, otherwise no deal.

When there is no agreement about a decision,

talks continue until an agreement is reached.

There are no rules "from the top down". A majority cannot force it's will to a minority. The whole idea of hierarchy is no longer valid and is replaced with a circle where everybody has their place.

2.3 Examples of prices and transactions

It takes some time getting used to the idea of the colours. Probably you are used to confound price in Euro with intrinsic value. That price gives a very simplified image of reality.

You will have to learn to think "product cycle". Is the cycle closed? How much energy is needed to make the product? How much energy is released in the decomposition after death of the product? How much energy is needed to start up the process of decomposition?

When you can answer these questions, or at least have a frame of reference, you can say something about "Value". In real life it could be that your community elects to adopt a demand and supply price model. Then the market decides, just like now. There are also tables available to give you a fairly accurate picture.

If you are used to haggling about a price, you can continue to do so within colourcash. It's up to you. The only thing that changes is that you use a currency unit of three instead of one. And, every currency has its own meaning, based on universal energy in sunflower oil. But later more about this. First a couple examples about the colours themselves.

The following examples are about the combination of colours which form a price of a product. You have the choice between several alternatives, each at its own price.

2.3.1 Example: Transport

To illustrate the use of colours, lets look at some examples of which colours are in use for what kind of products. Fi2rst, we look at several transport solutions: the bike, a car and the air plane.

Bicycle: Red and Green, the food you eat.

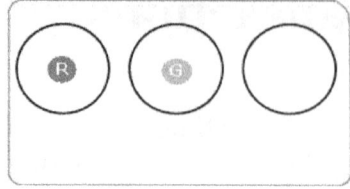

Figure 8: RGB Value of a bicycle trip

A car uses Red and Blue, and you use a lot more than a bicycle.

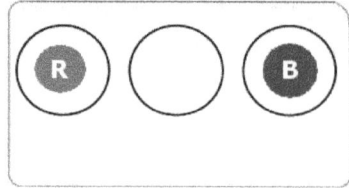

Figure 9: RGB Value of a car trip.

A plane uses the same , Red and Blue, but a lot more.

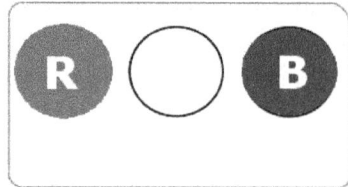

Figure 10: RGB Value of a short air plane trip.

As a matter of fact a bicycle is a lot cheaper, see table 1 on page 29. There you can find the correct ratios.

2.3.2 Example of colours: The bottle

You set up a business selling home made sunflower, corn and hemp oil. Which type of bottle are you going to use?

A cardboard, milk carton style packaging. Plastic and cardboard are fused together, so they can't be recycled any more.

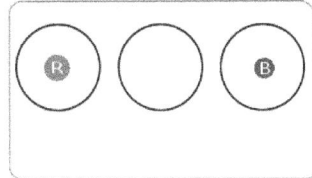

Figure 11: RGB Value of a milk cardboard.

New glass contains max. 50% recycled glass. So a max. recycling of 50%, but it has a higher energy cost because of recycling.

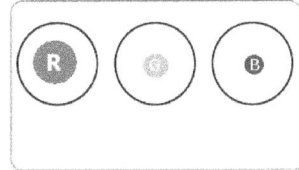

Figure 12: RGB Value of a glass bottle.

An earthenware jug is heavier and takes more energy to make. But the earth can be recycled.

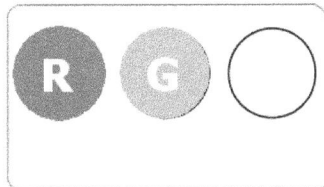

Figure 13: RGB Value of an earthenware jug.

These examples deliberately don't use numbers. They're just used to give you a feeling for the colours. The size of the circles is enough of an indicator.

2.3.3 Energy yield of foodstuffs

To make it more concrete, there is a list of commonly used foodstuffs and their nutritional value. This is the amount of energy released at combustion. This is a nice starting point to give you an idea of the amount of energy that was put into the foodstuffs in the first place to produce them. When a foodstuff costs more energy than it yields, it's not profitable. This doesn't mean it's bad, it's just not profitable. When in need, eliminate these non-profitable products first, and go on from there. In real life this will mean that prices of "animal" products like meat and fish are way too low. Also products like

---

Food return on Energy Investment

Today's food-industry requires lots of energy. Food actually is an energy container of solar energy, but agro-industry requires far more energy than just to grow plants. Transport, cooling, heating of greenhouses, industrial processing and the use of pesticides and artificial fertilizers require about 10 times the energy of the food it delivers.

A unit of meat and fish require about eight units of grains, which is used as nutrition for the cattle and cultivated fish populations.

In case of cattle / meat fed with industrially produced food, both multipliers... damn that's a lot!

To compensate for Energy Investment, use the following rules of thumb:

- agro-industrial food : Red x 10

- meat: Red x 10 x 8

- fish : Red x 8

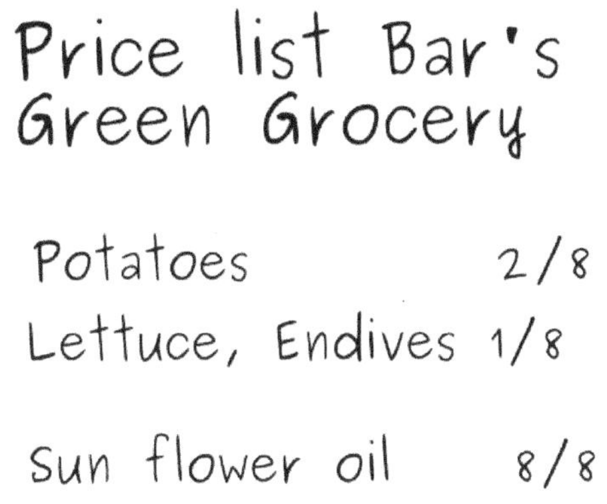

Figure 14: Price list of a Green Grocer. No need to even mention the weight.

cheese, butter and milk are higher, but to a lesser extreme. To calculate the exact value of the various products, you need to look at the costs, or, the food the animal needs. In the case of fish, you have to calculate the so-called by-catch. There is also a price to the fish used as fish feed in fish farming.

The energy in 100 grams of sunflower oil is used as an index.

1 Red = 3400 kJ = 810 kcal

We chose to divide the scale in 8 parts, strawberries and peppers fall into scale 1 and the oil itself into scale 8. Also, the lowest and highest partitions are only half as high as the others, so the divisions 2 till 7 each represent 1/7 of the index, 485 kJ.

This is an arbitrary scale, but the same goes for scales to measure wind (Beaufort), earthquakes (Richter) and temperature (Celsius, Fahrenheit, Kelvin). Eventually the main thing is that the scale is workable. Time will tell.

Table 2 gives a clean division: similar products are in the same scale. Notice that things like transport, packaging or processing aren't included. We only reckon the energy of the product itself.

Scale	From [1] (kJ)	To (kJ)	Examples of foodstuffs. Energy within the product, nutritional value	RGB
1	0	242	apples, beer, beet, carrots, cauliflower, cabbage, cherries, cucumbers, peppers, pears, leek, rhubarb, radishes, strawberries,oranges, lettuce, string beans, spinach, sprouts, yoghurt, blackberries.	R1/8 G1 B0
2	243	728	Potatoes, bananas, peas, veal, chestnuts, chicken, eggs. whole milk, breast milk (human), cow's liver, fish.	R2/8 G1 B0
3	719	1214	Brown bread, brown beans, green peas, boiled ham , raisin bread, fries, whole meal bread, rye bread, ice cream, beef, pork, white bread	R3/8 G1 B0
4	1215	1700	Rusk, brown sugar, honey, cheese (mature, whole milk) liver sausage, rice, whipped cream, white	R4/8 G1 B0
5	1701	2185	-	
6	2186	2671	hazelnuts, milk chocolate (bar), peanuts, sunflower seeds.	R6/8 G1 B0
7	2672	3157	margarine, butter.	R 7/8 G1 B0
8	3158	3400	Tallow, bacon, sunflower oil.	R 1 G1 B0

Table 2: Examples of energy contained in food.

Most other kinds of kitchen oil e.g. olive oil and sesame seed oil, contain a bit more or less energy depending on the brand. Olive oil and sesame seed oil belong to category 9, because of their powerful 3700 kJ per 100 grams. Well over sunflower oil. They will be slightly more expensive in the shops. No doubt

1   Per 100 gram product. source: BINAS

they were more energy-expensive at production.

2.3.4 Energy yield of fuels

 Mineral oil and her derivatives like petrol and petroleum are part of our "normality" as fuels.

Unlike with food we will never know exactly how much energy it took to make many of these energy sources. Especially mineral products that are mined. Luckily the Binas - the Dutch high school scientific reference handbook - tables just give the exact combustion per litre value. The values from Binas have been recomputed to contain the energy per 100 gram. This is unusual when we deal with petrol, because now we are used to measure this fuel in litres, but you'll get used to it. Anyway, a lot of substances are 1 litre /1 kilo. Check it out for yourself in Table 3.

Scale	Examples of fuels. Energy return on combustion	Red
1		
2		
3	peat	R 3/8
4	wood	R 4/8
5	brown coal, methylated spirits	R 5/8
6	methanol, ethanol	R 6/8
7	alcohol, hard coal	R 7/8
8	*Sunflower Oil Reference Index*	R 8/8
9	natural gas	R 9/8
10	petrol, butane , oil (reference TOE)	R 10/8
11	acetylene, propane	R 11/8
Scale	Examples of fuels. Energy return on combustion	Red
12		
13		
14	diesel	R 14/8

Table 3: Examples of energy contained in fuels

It's remarkable how our daily fuels are a lot more energy-rich than sunflower oil, from scale 8 and up. We also see that the Swedes who commonly use spirit to cook when on holiday have to wait a lot longer for their food to be cooked. The popular Campingaz brand sells butane and nowadays the more powerful propane. Truck drivers and sales representatives will no doubt recognize the power of diesel. The huge potential of this fuel is almost double (14/8) of sunflower oil. In Euro, diesel is cheaper than Diesel, but in Red diesel is 40% more expensive.

Energy return on Investment (EROI)

Oil derivatives such as petrol and diesel also require lots of energy during production. This energy is used for exploration, extraction, transport, storage and processing. Due to Peak Oil, oil extraction has become more energy consuming. The EROI has decreased from 1:100 to 1:12 - 1:18 today. To compensate for EROI in the price, we can use this rule of thumb: simply add R 1/8.

Table 3 only shows an R-value. In contrast to food these fuels aren't really recyclable, considering you burn fuel. When we talk about bio-ethanol, it's recyclable and so Green 1. The other fuels are in principle Blue 1.

The value for oil in scale 10 is based on another standard, Ton of Oil Equivalent (TOE). This is an industry standard of 41,868 GJ per ton. This is the same as 1.23 times the energy in 1000 kilograms of sunflower oil. The difference is only 2%. That means that all products should be in the same scale as in the sunflower oil index.

2.3.5 Is 1/8 not accurate enough?

A more accurate scale is possible to split up the vegetables in scale 1 into vegetables 1/8, 1/16 and 1/32 of the index. It remains to be seen if this makes sense or only contributes to make things more complex. For now we didn't go into that question. We're already happy when we can illustrate the difference between Red, Blue and Green and are able to separate energy-rich from energy-poor products. Splitting energy poor products from even more energy-poor products doesn't seem relevant at this stage, and so it's not worth the trouble of calculating.

2.3.6 At the shops

The easiest example is farmer's market, a shop that doesn't only sell but cultivated what they sell. Only the Red/Green ratio is indicated, because everybody understands that the cultivation is 100% organic and so there is no Blue. We don't even need a "per pound" indication either.

Potatoes 2/8 means that per 100 grams of potatoes 2/8 Red is asked. A kilo of potatoes is therefore 20/8 Red and 10 Green. We only reckon in eights, to make it as easy as possible. Also, there is no need to indicate the weight (as in 2 Euro/kilo) , because the weight and the Green value are 1/1, and you extrapolate the Red value from the ratio.

A customer buys a given amount of products. In a note book the cashier makes the following calculations:

Product	R/G Price	R	G	B
5 kg potatoes	2/8	100/8	50	0
600 g endives	1/8	6/ 8	6	0
1 litre sunflower oil (1l = 1kg)	8/8	80/8	10	0
Total		186/8 = 23	66	0

Table 4: Computing prices at the counter.

There are a few things that come to light in this price list. Let's take a look at the products. The potatoes contain almost the same amount as the oil. Remarkable because the potatoes are 5x as heavy. A look at the price explains a lot.

We continue the calculation in eights until the total. This reduces the margin of error, and it doesn't take that much extra effort. It's relatively easy to learn to divide by eight, because if you can't do it, just divide by two, again by two and again by two.

A lot of people try to add up the totals in RGB to each other: R 23 + G 66 = 89 RGB. *That would really be handy, because it makes it easy to convert back to the Euro.* But the question remains... Can you deduce the amount of energy and recyclables in 89 RGB? Try it!

2.3.7 Payment

Now we know the total, R 23 G 66 B 0, payment can be made in a number of ways. Keeping an "account" is the easiest when dealing with regular customers. The customer can sign off for their acquisitions. At the end of the month or week the shopkeeper enters all the data into an on-line payment system, and amounts are transferred from one account to another.

Another possibility is payment "in cash". Then you need printed paper money or coins. That's cumbersome, so we could use cheques. This cheque is also "cashed in" via the on-line payment system. This is handy if we're dealing with a "once-off" customer who doesn't have an account. In the chapter about money creation we will come back to this.

Playing shop is relatively easy. When we look at entire industries, it gets quite a bit more complicated. Soon have to deal with several products "that belong to each other" and the story gets very complex.

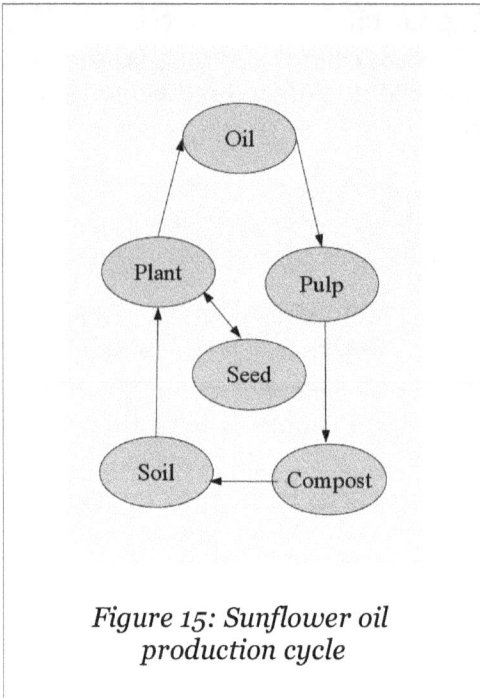

Figure 15: Sunflower oil production cycle

2.3.8 The production cycle of vegetable oil

Economists dream in their economy often about macro, meso and micro-level. We already saw some examples of micro level, such as the choice of mode of transport and the price list at the farmer's market. The next level is the meso-level, the production cycle. On this level, companies work together to make the final product. In our example we deal with the production of vegetable oil.

We are in charge of one hectare of land, that is, 100 x 100 meter. We presume that, in our region, we can harvest 1400 kg of seeds. 1400 kg = R 14.000. One tenth will be used as seed stock in the following season.

The pips contain, according to the table, 2435 kJ per 100 grams, or 6/8.

We deliver mostly to a number of presses in our vicinity. We could do it ourselves, but for argument's sake, we sell it.

Together with our clients we agree to market the oil at R1 G1 B0 standard price. We cultivate organically, without the use of pesticides or chemical fertilizer. Our sunflowers yield enough for a yield of about 600 kg oil. At the agreed price that means R6000 G6000 B0, the rest is pulp and is recycled. This way we close the circle of materials and the Green currency.

The consumer occupies an important role in the cycle, because they buy and use the oil. In principle we only reckon with compost toilets, so none of that matter is lost and the consumer gets the paid green back in their wallets. If you think this ludicrous, you probably aren't aware of any possible problem with the nutritional value of our daily food.

Market prices are set by taking standards in price of the final product or the separate products. When an energetic value is unknown, the people involved have to agree to a reasonable division. Only when they agree, they can market a product together.

2.4 Money creation

A lot of parts of Macro economy become irrelevant in colourcash. The whole phenomenon should be rewritten. If you have a background in economy, I must congratulate you with your persistence reading this book as far as you have. From here on it gets worse, we're going to talk about money creation and money destruction.

A lot of people think money begins its life when they take it out of the ATM. Taking money from your bank account does not mean you "create" money in that instant. That money was already sitting on your account, so it originated earlier. But the money on your account has to come from somewhere? "My employer!" I can hear you think. And he gets it from his customers. Right. But where did money start its voyage? When did it appear on someone's account for the first time?

According to the Belgian-American professor Bernard Lietaer [liet2001] there are two possibilities. The first is the case of mutual credit.

2.4.1 Mutual Credit

When using mutual credit one person sells something to another, both bank account balances change. The total of all bank balances should be zero, in principle. There is no money in circulation; all the money is virtual, on paper or electronic. This system can easily be introduced on the Internet, and no doubt it will be.

We used the same system in the student flat where I lived for years. Everybody had their own account at the kitty. When you paid for bread or dinner, you wrote it down. You could also settle internal payments. Say that you borrowed 10Euros, you add it to the lender and subtract it from the borrower. This way the total end balance is not influenced by an internal settlement.

This method of mutual credit works very well. It should be possible to introduce this scheme on the Internet. Preparations to this end have started. Keep up to date via www.colourcash.org , our official English website.

2.4.2 Printed money

The other way starts with printing paper money and minting coins. An old problem in the Capitalist (black and white) economy is that when there is too much money on the market, it loses its value. This problem is non-existent in colourcash. You plan ahead based on the expected harvest and use that base to create money.

We use the traditional Mayan "Day outside of time": July 25 for money creation and destruction. This day our Sun is aligned with Sirius in Orion, our nearest point of calibration in the time-space dimension. Our stellar clock heralds the new year then. You've already understood... Together with the money system we also reform the calendar, but that choice is up to you. If you use the Gregorian Calendar, I think December 31 or June 30 are good dates.

July 25 the sunflowers are nearly ripe to be harvested and the farmer says: I think I can harvest 1400 kilos of seed this year. He agrees with the oilman and they set the price for 60 kilograms of oil at R6000 G6000 B0. In the same way baker miller or other farmers make price agreements about the cycles of grain and bread. Also the hop farmer and brewer make their planning.

The evening of 25/7 all representatives of all (production)cycles gather and count their blessings. That moment, form the bottom up, the planning of our village tor the following year is born.

Product	R	G	B
oil (**)	6.000	6.000	0
bread (*)	12.000	16.000	0
beer (*)	6.000	8.000	0
+	=		
	24.000	30.000	0

Table 5: Planned turnover of three products in a village.
(*) fictional number. (**) real number

The total turnover of the village is R24000 G3000 B0.

The village knows, approximately, what the total profit will be.

The incomes of all villagers will be precisely R 24000 Organic waste, water and other recyclers will have to process G 30000. In this case a community of ten villagers can generate annual income of R 2400 per person or 6 per day. Not a lot, but the available space is limited!

All other products are procured in neighbouring villages that each have their own speciality, apart from bread. E.g. horses, cheese, wine, whatever...

The villagers will obviously spend money. They assume that all the money will circulate 1 x month, so they need to create R2000 G2250 B0. They can either print the money using a basic printer or keep a simple bookkeeping that regulates the mutual credit. This administration can be kept on-line, using the internet, or on paper in a ledger.

All this trouble to make a plan for the whole village, is something they'd probably have to do anyway. In an Every man for himself economy, you don't have to reckon with your neighbour, but if you have to survive with what you can produce...

2.4.3 Limitations

The given example is rather limited, so let's examine them closer.

You will have noticed that the earnings are rather slim for farmer and oilman, but that is what you get when you only have 0.5 hectare per person. A fair earth-share according to the Ecological Footprint is around 1.8 hectares per person. In Holland the space used amounts to 3. We are used to too much luxury, we use up more than we should. The inventors of the Footprint try to make you feel guilty in order to get you to take some positive action. Ask yourself... What happens if all those football fields (out there in Elsewhereia) I need to keep up my lifestyle, all of a sudden no longer are available to me?

The given examples was deliberately disfigured, because the Blue-part is zero, and that is an unrealistic situation. The goal of the example was to show how to determine the cost price of products in Red. There are no typical Blue products, like natural gas, plastics and imported goods, in the example.

So this example is far from ideal. No problem. I gladly leave other examples to reality when they come up. The examples I can think up behind my desk, with a bit of help of tables and publications, can't attain the same level of accuracy of real specialists in the field. Probably my figures using sunflower oil are so inaccurate that a specialist isn't convinced. That's why I invite the experts to determine the value.

2.4.4 In a nutshell

To determine a "right" price, you're dealing with:

- your partners with whom you make up your production cycle. The aim is to search for the true value, not to fill your pockets with money.

- The energy (kJ) in the product , or the energy released at the end of the product life. Food, composting, incineration).

- The amount of material (kg) that is to be recycled, according to agreements with your partners..

- the amount of material that can't be recycled because no partners could be found.

I took tables for the nutritional value of food, because in principle the amount of energy released by the food in the body is the same as the amount of energy stored by the plant and the sun. There is also a big part of the plant that isn't eaten, e.g. leaves and stems of the potatoes plant. These parts are not counted in the values, this energy will be released on the compost heap as heat. If you can use that energy, you can include it in your "wealth". If you don't do anything with it, this energy is lost. Not really lost, because this heat is needed for composting!

Cultivating plants and transport also take energy. This wasn't incorporated. Depending on the agricultural techniques used and the way of life, the production costs can vary between 1/10 of the yield to 10x the yield. This is the difference between our way of life and permaculture. Before we can understand this balanced way of life, we have to investigate our own way of life.

2.4.5 The Blue heap

In our modern way of life we produce an enormous amount of waste. Either at production level, transport or electricity.

Say that you, as entrepreneur, press sunflower oil. To press this oil you have to buy expensive filters from France, because that's where the best filters come from. You need new filters every year because the plastic gets torn. The factory doesn't take the used filters back. So you throw them on a heap in the back of your plot. You send the manufacturer an angry letter because you don't know what to do with all those heaps of used filters. The only alternative is to sell them to the destructor of the incinerator , who will give you Blue in return.

The pile keeps growing. The stuff isn't degradable. The acquisition of these filters are beginning to weigh on your budget. Because, even though the filters aren't re-usable, the factory wants to be paid in Green. Their reasoning is: We buy our raw materials with Green, so we want Green for our products, otherwise our books don't balance.

This seems a ridiculous example, but any example where the end-product can't be recycled, will give you the same result, the same problem. The producer looks at what he buys and the buyer looks at his waste, and they both say THEY are right. But who is right?

It's not about the finding the person breaks
the cycle, it's about who restores it.

There is no "destructor" who buys the stuff for Green. The manufacturer can't or doesn't want to buy it back for Green. Anybody who buys the product, chooses to go down a road without escape.

So we'll have to count Blue. Whoever handles the raw materials thus that they can't be recycled, is heaping up Blue.

This Blue heap means that you buy in Green, but sell in Blue. Inevitably you'll get a shortage of Green, and indeed, a surplus of Blue. Whoever chooses to stop recycling, gets stuck with Blue. What are the consequences?

You can't go bankrupt in colourcash, and you can borrow Green somewhere. But your books are available for whoever wants to see them via the Internet. You will be under moral obligation to produce more environmentally friendly. If you are prepared to improve your production processes, slowly but surely, the community will be prepared to fund you. The Blue you still produce is forgiven, but if you don't show any goodwill to improve this situation, people will criticize you.

Anybody with a bit of entrepreneurial feeling takes a look at their business.
Oh Oh, I'm stuck with a bunch of products for which I can't find a destructor. Before long I'll turn into a Smurf!

That is why it's important to find each other. Constructor,

consumer and destructor. It's possible to live like this, because people have been doing it for years. It's called permaculture.

2.5 Permaculture

Permaculture is a contraction of permanent and agriculture. It means that the land should always be under cultivation and never left fallow. But it goes a lot further than that.

Permaculture is Bill Mollison's brain child. In permaculture, the designer's manual, his explanation is thus:

> He realized the forest was constantly producing, only using energy of the sunlight and a fertile soil. There is no extra energy input and no waste.

This system thought is also at the base of the Enschede Vision (see page 25), a document written by young engineers that was the base for the design of colourcash. Permaculture and colourcash have the seam ideological base. We can say that this type of agriculture and the economic system were made for each other. Optimal energy use, closed cycles and co-operation are principles both share.

Unfortunately, this is not a permaculture book. But, in order to achieve colourcash, I believe, we need to start off with permaculture. You can start in the garden or the balcony and grow some of your own food.

Isn't that strange? You have a book on economic theory and the author wants you to start growing a kitchen garden. Why not stick to the subject and tell us what economic initiatives are available just like this one?

Part 3. Close Neighbours

3.1 Colourcash and other alternatives

Over the last 20 years a lot of people have gotten interested in the same subject I got interested in. They have a number of reasons for this, but all can be categorized, just like colourcash, under sustainable development. As far as I know they all chose for one of these two:

- a slight change in the Euro/dollar system

- a new system based on a single unit of value

Figure 16 shows the possibilities of a one-dimensional "colour" space. It's a bit like this: the higher, the better, the richer, the whiter. All values are crushed to fit this simple scale. No matter whether it indicates dollar, Euro, kilojoules (energy) or kilograms (mass).

Figure 17: Colour space of red, green and blue. Not visisble in the black & white version of your book. Please download the free PDF version on colourcash.org

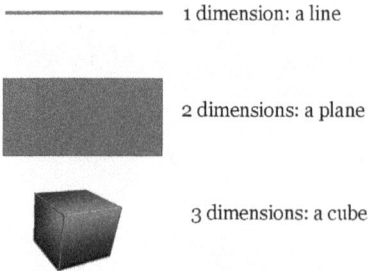

Figure 16: Colour space of white.

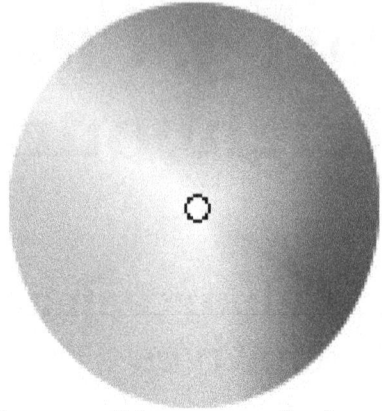

1 dimension: a line

2 dimensions: a plane

3 dimensions: a cube

Figure 18: Dimensions of space

3.2 Dimensions

Colourcash is unique in that we use a triple unit or trinity to describe "value", while other systems are stuck in the limitations of a one-dimension space. Figure 17 demonstrates the possibilities of a trinity. Using the three basic colours we can obtain almost all visible colours. There is no up or down, better or worse, it's about balance.

We've come across the notion "dimension" a couple of times now. It's important to specify what exactly we mean by it. This is the core of the idea after all.

Every mathematical, logical or physical space consists of one or more dimensions. A line, the simplest mathematical space, is one-dimensional. The economic world we live in today uses this very simple space. It reckons only more or less or the same amount.

The next shape is a plane. This is a 2-dimensional space. We reckon with Length and Width. Every spot in the plane can be indicated by two numbers, latitude and longitude.

The cube is a 3 dimensional object. Except length and width there is height, now you need three numbers to indicate any given point in the cube. This 3-dimensional space we use in colourcash because we think of the colours Red, Green and Blue as dimensions.

Multi dimensional spaces are possible in principle. Physicists proved that our world contains more than 10 dimensions. We only recognize three. These three dimensions do not have a specific "order of importance". When you're in a cube, it doesn't matter whether length was considered number 1 or number 2 by someone who lived in a lower dimensional system. Talking about a "6th" or "9th" dimension doesn't make any sense really. You could say "this happens in a "9 dimensional" space.

After this rather theoretical part about dimensions, we can return to our subject, alternative money systems.

3.3 Proposals for adapting the present system

First we will deal with proposals that imply a change in some way to the present money system.

3.3.1 Deposit (e.g. for glass bottles)

Getting money back for re-usable bottles. It seems as old as the hills and logical. But it meant a real change in the money system. When you buy a bottle of, say Cola, you buy the right to bring back that bottle. It's a kind of double production chain, one for the cola, and one for the bottle. 2006 saw the abolishing of the compulsory deposit for plastic bottles. The coming years we will see a growing number of single use plastic bottles.

The deposit-regulations and abolishing the obligation of same is an indicator of how much recycling and collection of valuable re-usable raw materials depend on the political agenda of the day. Recycling is not an intrinsic part of the system. Adding it afterwards means that it can be taken away again. We see the same problem with ecotax.

3.3.2 Ecotax

Ecotax is a tax on environmentally unfriendly products and a lower tax on environmentally safer alternatives. We could attack the pollution production itself. A second option is to tax the use of (non sustainable) energy.

Ecotax was never implemented on a large scale. Earlier on in this book I already told you why. The money system is so unstable that it would collapse when fundamentally changed. Nobody dares to do this.

Say there was an ecotax. Then there are still some serious limitations. At the end of each production cycle there's a choice between recycling and "Blue heaping". When dealing with synthetic materials that are difficult to destruct, you are using extra energy. These costs are included in the price. But what is the manufacturer to do now? What will the consumer do? Will there be recycling or does one take the ecotax in their stride and move on? Or do we leave the waste where it is and pay a tax on it? What is the correct ecotax on two fundamentally different things like "energy" and "material"? They are both reckoned in the same money and this makes it difficult to get to a reasonable rapport.

ecotax doesn't change anything to the structure of the economy. On side of the economic (down) stream the waste heap continues to grow, only now tempered a bit with an extra tax, which will give new wind to a second "recycling" industry.

The inherent duality between energy and matter is not included in the ecotax concept. Yet ecotax is a good temporary measure for a society on its way to Sustainable Development. ecotax is useful as an alarm clock, to wake up entrepreneurs in money-land and make them aware of the need to innovate.

3.3.3 Organic, biologic, Bio-Dynamic and Eko

The last thirty years a lot of very important work has been done in the field of agriculture, no poison, no chemical fertilizer and animal-friendly. Innumerable labels and stamps are the result. Except for the taste there is no real other way to recognize the trustworthy product.

A big disadvantage of these organic or bio-products is their price. For a jar of bio apple sauce you'll pay 50 cents more, even if the bio apples themselves only cost about 1 cent more that conventional, non-bio apples.

The reason for this is the middle man. Everywhere there is price-pressure and for a small party such as a bio-apple sauce brand the jar manufacturer charges a higher price. And hat results in a higher price at the shops. As a matter of fact, all production steps are more expensive, because the small producer has a lower turnover than the big producer. This show that the bio-sector has to row against the flow. And their means are their enthusiasm, ideals, love for the trade and know-how.

The consumer isn't really convinced yet about the "why" of bio-products. Although the price is higher, a bio product is not easily recognizable under the harsh TL light of the supermarket. Only when tasting the product you might notice a difference in taste. Just like Fair Trade, Bio only has a market share of about 5%, and that's a pity.

Colourcash offers opportunities for both producers and consumers of bio-products. All of a sudden your conscious Bio buying is rewarded with a better, more balanced price. Growers who use poison have to reckon Blue; Growers who use chemical fertilizer have to reckon with a big rise in Red-buying. Production of artificial fertilizer uses a lot of energy.

The consumer now sees the difference between a bio apple and a conventional apple by the price. A bio apple doesn't have a Blue price. Because the venom is also on the apples, the grower of conventional apples has to reckon Blue to his customers. And the amount of Blue shows us the concentrations. And if he doesn't charge Blue?

We will know immediately because of the transparency of the bookkeeping we will find out about the grower buying venom. It's not the end of the world, but now at least we know.

The first colourcash users will be bio farmers and horticulturists. The network offers contacts with clients and long-term business relations. And also respect for the know-how acquired!

3.3.3 Tobin Tax

Tobin Tax was named after the economist James Tobin. He came up with the idea that took his name, after President Nixon had to abandon the Bretton Woods system.

Tobin tax is levied on money that crosses the National Borders. This levy can be low or high, between 0,05 and 1,1 percent. This tax has as a goal to put a stop on money speculation. Speculation is the cause for exchange rate fluctuations on the money market because it moves big amounts of money from one country to another, obviously done electronically.

Tobin-tax hasn't really broken through in real life because the people in favour of this tax are waiting what others will do with it. According to Wikipedia, President Chaves of Venezuela is about to introduce this tax.

Tobin tax is nothing more than a plaster really. It only taxes the moving of money where there is no movement of goods. The amount of money has to be in relation with its goods. So if you move money, but not goods, you unbalance both countries.

Colourcash makes it principally impossible to move money without moving goods.

The Dutch know all about import. One of our strong points is the import of animal feed. This can be done, but at what price? colourcash dictates that everything should be recycled in order not to end up with a Blue heap. This specific Blue heap we call the manure surplus. Or the Dutch livestock holder becomes a smurf, or he exports goods of similar value like compost to the countries of origin.

3.3.4 Sustainable, ethical and responsible investments, people - planet -profit, environmental performance

This also deserves support. People try to reach goals and improve situations from an existing situation. Hurray!

Personally I am rather sceptical towards a big number of so-called responsible enterprises. The Telecom sector is very popular in this view. But what about GSM/UMTS- radiation? Proof that this radiation is harmless has never been provided, and that is a factor for responsible enterprises. Coltan-mining in Africa, you say?

Another example is Vopak, the largest Dutch oil storage company. I am very happy they are concerned about safety, because I live within a 5 km radius of hundreds of their oil tanks in the Rhine estuary. But in a sustainable world there is no space for a giant oil industry as an IV for car-it is. So how can this company be sustainable?

What matters is finding out how "sustainable" we really are at the moment. This is possible by publishing company-data. This in turn is possible by using the RGB trinity.

The experience gained with publishing sustainability data will be very necessary in the development of colourcash. The closure of the life cycle of their products and the minimal energy use are daunting challenges for these companies- And should they find out that the Euro economy isn't profitable, they can still join colourcash.

3.3.5 Internalization

"Internalizing" environmental costs is very popular with

scientists and politicians alike. E.g. when a car drives, burned fuel is dumped as "air" (vapour, carbon dioxide, numerous poisonous gases) via the exhaust. This is pollution. This pollution isn't included in the price of the car or the fuel, they say, so if you do calculate these costs in the price, you internalize these costs so to speak, you discourage pollution.

On itself "internalization" is a good policy. Nothing wrong with it.

The only thing I don't like is the word "internalization" itself. If you look at the product from it "cycle" point of view, petrol is still petrol after combustion, albeit in a different phase. It's not a surprise you are going to burn that petrol, is it? So how can you call this pollution "external". Typically "production chain" thinking!

3.3.6 Micro-credit

Micro-credit is lending small amounts of money to poor people via local banks and support groups. It turns out these "poor" people are just as reliable fro repaying their debts as the normal target groups of banks. In that sense there is no fundamental change, only the point of view of the market has shifted a bit.

The success of micro-credit shows that when "money-less" people get access to money, a whole new world is opened to them. I see this as encouragement for colourcash. This means that opportunities for people in one part of the world don't mean a greater lack of opportunities elsewhere. See "Interest" and the story of the last tenner on page 131.

3.4 Alternative Monetary Systems

The following systems are not an adaptation of the present system, but new creations. The main point they have in common with the present Euro-system is that they are based on single units. Black and White money, as we termed it. Bernard Lietaer calls them "complementary" monetary systems in his book The money of the Future [liet2001] because they are suitable as an add-on on the current money-system. They are mainly used in places with a high unemployment rate and economic misery. The number of systems globally is experiencing a "boom". In 1998 there were about 2000 systems in use. After the money-crisis the number of users in Latin America rose to millions of users.

I can only mention a number of examples, all of them taken from Bernard Lietaer's Money of the Future. [liet2001]

3.4.1 LETS

Lets is the best known alternative money system in Holland. The Local Exchange Trading System is made up of a network of shopkeepers that exchange goods and services based on mutual credits. The "credit" of the buyer gets lower; the seller's credit goes up. At every acquisition the buyer signs a check which is then processed in a central administration. The total credit remains zero in principle, if it weren't for the "wages" of the administration.

In Dutch cities this LETS system is used for acquisitions of services and used goods. There is hardly any offer of new goods or things like vegetables. Using colourcash language, I would say Red and Blue are for sale, but there is no Green. This is logical. We are talking about city dwellers and they may think there is no space available for a vegetable garden.

Up to now the LETS users haven't been interested in colourcash because they feel it too difficult. Also the change over is difficult. What is the value of your "stars" in colourcash? It's also very hard to convince groups of hundreds of users. In short, LETS users will keep on using LETS as long as they are happy with it.

The big difference between colourcash and LETS is that colourcash tries to include the vegetable and fruit production. Green is nice and local production doesn't take a lot of energy. When city dwellers are supported by villagers from surrounding villages there is space for an interesting network. City dwellers provide labour force and customers, villagers the land and know-how. An excellent combination.

3.4.2 Time-is-money

A lot of systems use time as a value base. Ithaca-hours en Time-dollars are a few examples. Time you spend on helping someone else can be used to be helped yourself. Or you could use that time for your grandmother who lives 100 miles away, if they accept the token. This way you can help your granny, whom you hardly get to see, by helping your neighbour.

The time-dollar system is popular with retirement homes. The people there don't get just more help, but get to know each other better and a closer knit community is the result. People even gained health. Insurer Elderplan from Brooklyn, New York, accepts the time-dollar as payment for up to 25% of the premium. They hire "staff" with the hours they accumulate.

Again we see that these systems are most successful in places where there is the least money circulation. However, this isn't always the case. Let's go to Switzerland.

3.4.4 Wir

In 1934 WIR (or "We) was founded in Zurich. It's the oldest

still working complementary money system in the Western World. Today it has about 60 000 participants. The system is special in that it encourages spending a lot of "Wir" before making any. The contrast with the Swiss Frank is huge... there you are supposed to earn first and then keep spending to a minimum.

3.4.5 Terra

The Terra is Lietaer's own attempt to set a unit of value, as a reference for other systems so international "inter-system" trade is possible. The Terra is such a unit and is represented, as an example, like this:

$$1 \; terra =$$
$$1/10 \; barrel \; (= 15,9 \; litre) \; oil +$$
$$1 \; bushel \; (= 21,7 \; kg) \; wheat +$$
$$2 \; pound \; (= 908 \; gram) \; copper +$$
$$1/100 \; ounce \; (= 0,31 \; gram) \; gold$$

By using a reference to concrete, physical goods it's always possible to measure the value of your local unit to the Terra. The same trick is applied in colourcash, except that we chose there for another unit, Sunflower oil, its energy and its mass.

The rest of the differences is in the details. Of course colourcash uses a trinity and not a single unit. But, by taking natural units of mass and energy it's possible to get a fairly accurate estimate of the value. The Terra is coupled to market prices.

A big similarity between Terra and colourcash is that they were invented behind a computer screen and haven't seen the real test in practice.

The test in practice is that there is no International Standard of Value, and there hasn't been since 1972. Yet you can use your credit card to make payments over the Internet in countries on the other side of the globe! I expect that the physical "coverage" of units like the Terra and colourcash are more stable in the long run than the currencies we use now. And this is not very important for the bread you will buy tomorrow, but it is vital for the pension you will get in 2036. This kind of practical subjects we deal with in the next part.

Part 4. A-Z

In this final part we deal with everyday economic topics and relationships to economy, man and environment. In colourcash they are connected and cannot be separated.!

Alcohol

Alcohol is found in numerous drinks that are famous for their region of origin. The last years this phenomenon has gotten somewhat less since wine from Chile and South Africa can compete with say French wine. I expect the number of vineyards to grow in Holland since wine was produced here many years ago. I also think traditional alcoholic drinks will have a revival. Holland will become a gin country again. Beer is an Eastern European product that is popular now but can be replaced by original varieties such as fruit beer and beer with spontaneous fermentation.

Allowances, well fare

Welfare is needed for anyone who doesn't have an own income or whose own income is not sufficient to live on. It is provided by the State, a party that is used to shifting money from one side to another.

In colourcash, it's different. It depends on what you decide in your Sovereign Entity, your village. You could opt for a base income, the amount of which is easily calculated on the expected income (harvest) divided by the number of inhabitants. You could opt for the "every man for himself" system and supplement any shortages. Which groups are we talking about?

There are children, the elderly, sick people, the unemployed (?). Then there are homeless people, illegal aliens, orphans, psychiatric patients, you name it. We are used to "box" people, but they're all people. When they need an allowance, or more in general, help this is because they can't look after themselves. In that sense everyone is equal, nobody can live without their neighbour.

It's a bit patronizing to say; you're older/younger than * and so * you need this and that. No, Every person is unique and who can't hack it, needs a helping hand. Not based on statistics or other external numbers. In addition: because of aging population we can't afford so many elderly people just "hanging on".

Art

Art is the product of personal inner process. I myself never really understood art, but I do know that the first to quickly understand my lectures on colourcash are usually artists. I don't have to explain them anything about colours for a start. Then they are locked, just like many other people, locked in an impossible position where on the one hand they have to make money and on the other hand want to make art. They also know everything about the value of money, or read Art: The artist often earns less than their agent or customer. The customers are prospectors, whilst the customers of farmers (who are also locked in the same position) are whole sale concerns and anonymous customers at an auction. All potatoes are the same (in a manner of speaking) but Art is a unique product.

However, Art is mainly a "Red" product within colourcash. Every painting has its weight. The Artist makes his money in "Red". If he's a painter and he works with natural paints, then Green also comes into play, if the paints are synthetic, then Blue comes into play. For the raw materials time will tell whether they are Green or Blue.

Automation

Fewer and fewer people actually "work". Those who do control computers. (Not considering the computers that control people). A lot of the work can be done by human beings, but is done by robots.

The balance shifts. Consider that a human being uses the same energy as a robot or a computer. We already have the people. The robots and computers we can switch off. So in the end it's more energy efficient to do it with people, when possible and desirable.

In principle this rule is already applied, but people let themselves be used in the competition between two rival companies. This point becomes moot. The real energy use will be registered as Red.

Banking

A bank is a central organization that supplied financial services. There are all kinds of banks, big and small. Normally their goal is to fill the pockets of the stock holders (which normally are already full), creating a power base for the State or supplying services to members. This can be seen with co-operative banks, even though even they have commercial interests.

With colourcash and its complimentary programs it becomes possible to play "bank" in your own community. The banking trade no longer has any secrets with the abolishing of interest and inflation.

For existing banks it's possible to offer services in colourcash. This is possible for everybody, also for banks. Just like everybody they are faced with the challenge... Or will they turn their face away?

Bankruptcy

The biggest fear in a Capitalist world is bankruptcy. If your debt is too high and you don't have any future prospects all your possessions are sold for twice nothing to the bank. Main cause: Interest. Secondary cause, a system where competition and "being the leader of the pack" are the main values. Third cause, Bad luck. It could happen to anyone. Only some of us get out of it a bit easier than others. Quite often you loose everything, your house, your job and often your wife. To prevent this bankruptcy you have to earn a lot of money and save a nice sum to boot.

Entire countries went bankrupt, bust. Lost everything. Everything sold for a song to foreign money men. Thanks to interest and greedy dictators. And thanks to Western banks who lent "our" money. And now we complain that poor countries have built up such a huge debt...

There is no such thing as bankruptcy in colourcash. People work together, help each other. Loans are temporary and are written off on day "zero". When you go bust in a Black and White money system, you should be able to join group for your daily existence. The drain starts at the bottom, doesn't it?

Calendar

Why would one month be shorter than the other? Does the moon revolve slower around the earth in January, so it takes 31 days to complete an orbit instead of 28?

Neo, from the film The Matrix could have said about the Julian Calendar that it was just another system of control. The Julian Calendar cuts our link with our own natural rhythm. The only thing this calendar can do is to make Spring start when the calendar indicates March or April. Other than that the calendar doesn't make any sense at all, when you look at how time cycles work.

I don't mean to offer you a different kind of Calendar in this book, nor to explain why or how to use it. There are other books that do exactly that. But, realize that when you are on the path of colourcash, the calendar is one of these things you will come across.

Care

Children, elderly people and the sick need carers. In the old days the family took care of the day-to-day care. Everybody lived on a farm where there was enough space for everybody.

Nowadays we live in towns, in houses stuffed with things. The house I am living in offers room to some more people, if they are willing to sleep on the floor, but there is no more room for their "stuff". Benches, couches, book cases, desks and computer take up all the space. We also have our own cheese cutter, just like the neighbors. Nice, all that stuff just for one person...

The down side is that, because my girlfriend and I live

together, grandpa and grandma live elsewhere. Every generation in a separate "box", like Native Americans call the White man's houses. And yes, when granny needs help and grandpa can't do it, Home-care is called.

And their work is valuable, priceless, mostly done by women. But the baby boom of the 1947 generation is coming towards this Home-care like a Tsunami... in about 10 or 15 years. Reports predict that we'll have, depending on the region, between 20 and 25% of the generation in their twilight years. How many people are needed to care for them?

A lot of people put aside a nest egg or built up a pension. For all the others home-care will be too expensive. People will, more or less by necessity, take their parents in their homes. And look after them themselves.

In the Euro-economy manual labour is relatively expensive and so it's not profitable to get this kind of care. Colourcash is a lot more "economical" here because manual labour as a whole is a lot more economical. Bernard Lietaer tells us there is a Japanese complementary money system that is so successful that people prefer to be taken care of by someone paid for from within the system than someone paid for by the insurance company. The more personal approach and rapport are said to be the deciding factors.

Clothes

The famous heyday of the rulers of Florence had a direct link with the textile industry. The Dutch industrial era began with textile-manufactures and ended with their bankruptcy. Cheap, slavish Chinese labourers produce anything and everything, including textile.

Of course you don't keep clothes for trading but for wearing.

Clothes make the wearer. They can tell your surroundings who you are if you want them to.

Food might be the start of colourcash, but clothing is the decisive factor in colourcash. Valuable, pretty, nice, unique and affordable. A sign of independence and identity. Freedom. Emotion.

Just like in the late Middle Ages we will make everything with fibres that do well locally. Flax and Hemp for Holland. Cotton has a short fibre and it's difficult to spin it manually. Wool is rather obvious, but as supplementary material. Wool is too hot in the summer, isn't it?

The raw materials can be grown locally in the garden. Carpenters and toolmakers can still make the old machines, so tools like spinning wheels and looms will make a comeback. Save energy, push with your feet! And there's no need to go to the gym at night!

Once the fabrics are made, they can be transported and traded. On any given location they can be turned into clothing or other textile products like bed linen or curtains.

With a bit of luck local styles and even traditional dress will come back as well, but I won't make any predictions on that.

Communism

A lot of people I talked to think about Communism straight away. Here's the thought process.

Aha, you don't like Capitalism, the only alternative is Communism. So you are in favour of Communism. But Communism is bad and rotten and out of date to boot. Why don't you like Capitalism? It's the best alternative, isn't it?

This is a common thought process. It's usually very hard to explain that, form the point of view of colourcash, Capitalism and Communism are very similar, while colourcash is something completely different. Let's look at some characteristics.

Centralized and "Decentralized"

Capitalism and Communism are both centralized systems. There is a government which decides on emitting money. She makes the rules and regulations. How people get elected, democratically or not, I will leave out of discussion here. The contrast with colourcash is that the individual has to make room for a greater, collective "good". This is abused, time and again; by the way centralist power works.

Why does it say centralist and not just central? The idea behind it is that they still want you to believe Capitalism is a decentralized system, so not central. Decisions are taken from the top downward, and the top is always smaller than the base. In the end you have no say whatsoever, even if you have the right to vote. The decisions are not taken centrally, but by an executive, bureaucratic organization. (You can recognize these by their three or four letter abbreviations.) For every detail there is a separate authority" authorized" by the central authority. That is why I call this model Centralist and not Central.

Currency

Both Communism and Capitalism use the well known Black and White currencies, one dimensional system. In both cases the State is responsible for the emission of these currencies. The currency doesn't have any meaning as such and is linked to banking or legal constructions to the value of other "one dimensional" currencies.

Colourcash uses the "RGB", the three dimensional unit in

three colours, Red, Blue and Green. Anybody can set up a RGB banking system because its structure is decentralized. Whether and how to exchange with other systems is up to you, but it is possible in theory.

Ownership, State owned and responsibility.

Under the Communist System all property belonged to the state. The result was that John Doe was left with nothing and the Party members kept the luxury goods. Capitalism is the exact same, but it's called differently. Also here common folk have a lot less than rich people. Whether they have more in one system than in another is not important; in both systems people are confronted with a system they didn't choose. I was born in Holland but never nominated the European Central Bank.

The notion of Property is relative. People take something "in property" as long as it suits them, but after it is no longer useful, the product is exhausted, it is thrown away and the product is left in an unknown, undesired state. Dumped, let go, or burned.

In colourcash we turn everything upside down. We abandon the whole "Property" concept for "Responsibility". You no longer can throw away stuff when it suits you. And if you do, we'll see it on your coloured ledger.

Capitalism	Communism	Colourcash
Centralist	Centralist	De-central
Single-dimensional currency	Single-dimensional currency	Three dimensional currency
Private Property	State owned property	Responsibility

There you see, Communism has two basic parts in common with Capitalism. Colourcash is completely different. Conclusion: Capitalism and Communism resemble each other and colourcash doesn't resemble any of these.

Competition

Competition is a habit that stems from envy. People don't
want to see the light shine in the neighbour's garden, and that
is why we devise the most incredible tricks. I used to work in
schools. You weren't allowed to make bad news public,
because then you would endanger the good name of the
school, and the school wouldn't attract new students. Cause;
the other schools do the same. Schools compete with each
other, they have to.

There are more examples of competition that clearly useless.
But which competition we could define as "healthy" or "good"?
When a milk factory wants to sell milk on a market 1000
kilometres away from the factory, it's logical their prices will
be a lot higher than the prices of the local farmers.

In colourcash there is no competition. Yet there are price
differences. Far away is never advantageous for competition.

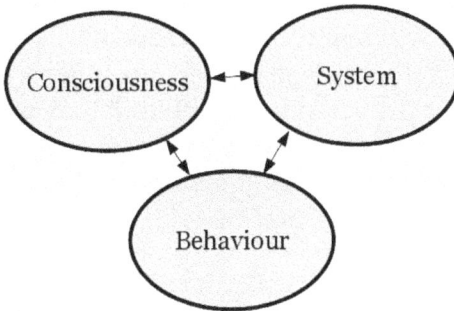

Figure 19: The relationships between consciousness, system and behaviour

Because its origins are too far away. Because of the higher fuel costs, based on energetic valued, dragging stuff over long distances is discouraged. Yet you could buy bananas. But do we really need to eat African beans every day?

Consciousness

It's not easy to describe the notion "Consciousness" objectively. One person sees the human being as a mechanical-chemical-biological being, the senses being the only connection between these components. Someone else, a growing part of the population- determines man as an "energy-being", and all people are connected by invisible "energy fields". This field is called "morphogenetic field" or "Collective consciousness/awareness" or simply "the field". This field would be the explanation why someone calls you just when you are thinking about them, or why someone turns their head when you look at them from behind. Whatever your interpretation, let me stick to "personal awareness".

The personal awareness is the way you look at the world. It has a direct influence on your behaviour. When you are aware that driving at high speed is dangerous, you will more than likely slow down. Or, when you know there are speed checks, you will drive a bit slower. It also works the other way round. When driving a car the world looks different than when riding a bike. The driver probably isn't aware it's more fun to use a bicycle. We keep this behaviour going via awareness. You don't know what you're missing.

In addition to Awareness or Consciousness and Behaviour, there is a third factor: the system. A system is a set of agreements, rules and procedures, plus an amount of information and goods that circle in it.

A pinball machine is a good example, with its lights, bumpers and counters. The ball goes round and round in the machine, from flipper to counter. And the counter on the display keeps the score. This is the system. The ball going round and round is our behaviour and the "consciousness or awareness" is the player. The fun, the excitement and the disappointment when the ball disappears...

System, consciousness and behaviour belong to and influence each other. The three represent a part of reality. The better they fit; I mean resemble each other, the more fun it is to play. It's called Coherence.

This book is about a system, colourcash. This is playing with colours and numbers. When we expand to mentality, co-operation, openness, transparency and the like, we are talking about typical elements of consciousness. What makes the player tick? We also hope to include behaviour in time. Then it will really work!

Death

Death is one of the last taboos in the Western World. Sex (who does it with the curtains open) and Money (who puts their tax return on their weblog?) are the other taboos. The taboo surrounding death is so strong Death hardly is mentioned at all.

Death isn't only important for personal life when loved ones or pets die. Without Death there wouldn't be space or food for new life.

Compare it to a parking lot. If it weren't for the junkyard, all the available parking spaces would be filled with old wrecks. There wouldn't be any space to park a new car. Death has been the territory of religion for centuries. Even the economy denies its existence. Death is kept out of principles or models. We change this in colourcash.

Debt

Debt only exists when possession exists. One man's debt is another man's wealth. However, we change the notion "possession" with "responsibility". We won't have any real debts any more. Because everything belongs to the Earth and so all is the Earth's possession. It's up to us to use everything in the cleanest possible way, so when we have to share, we'll just have to share. This is not some one's "debt".

That is why we zero all the balances on July 25. You start over again, every year again. Every year starts with zero. Nice and easy. You could keep a record on how you were doing on July 24 and use it as a reference for the following year. Excellent.

Next to this notion of "debt" in a financial sense, there is a moral "debt". Something went wrong somewhere and that is some one's fault, someone's debt. You get the Blame.

Right and Wrong are both sides of duality. Looking at it form a Trinity point of view you could say advantage and disadvantage. Even if some one is to blame for say, an atomic disaster the scale of the Tsjernobyl event then the advantage is that you learn it's better not to generate nuclear power.

The State is to blame for a lot of things that go wrong in the environment. In Holland a public servant can't be prosecuted (since the Pikmeer arrest) for crimes against the environment under his supervision. He was doing "his job". Because governments are collectively guilty of allowing crimes against the environment by industry, virtually all such crime goes unpunished. By dismantling the Government as State form this damage can be avoided.

Drugs & medication

There are two kinds of drugs, prescription and over the counter drugs. Because we will keep a sharp eye on our suppliers in colourcash, all drugs are considered alike.

The cycle is essential in colourcash. A lot of drugs are chemical in nature: developed in a laboratory, tested in clinics and produced in petrochemical factories. Their production requires a lot of energy, as is the research and development of new medication. Very Red drugs, so to speak.

The cycle of these substances is very unclear. We know that our tap water contains hormone-type substances. State secretary Van Geel expresses his concern in the magazine Elsevier (5 July 2005). British tap water contains Prozac, a powerful anti-depressant. Chemical drugs are nice examples of Blue. Unless it's proven the body destructs them into harmless, common substances, drugs are to be considered as 100% Blue. Their production is, per definition, out of balance. Because we've eliminated the "profit", there is more space for cures and medications based on simple preparations using herbs and plants.

This biological "medicine farming" is in colourcash, a lot more profitable. It will be relatively easy to cultivate medicinal plants yourself, and are relatively Red inexpensive. When we don't do any chemical tricks with the extracts, I assume, maybe not correctly, that these substances are absorbed in the cycle just as easily as we produce them. Whether they still end up on the Blue heap depends on our toilet habits, and not on the chemical nature of the substance.

Earth Charter

The Earth charter is a nice initiative where all kinds of principles are described and bundled. Its organization is the Steering Comity with its brass. Then there is an executive organization that prepares the documents and which provides training.

The problems the Earth Charter wants to resolve are the problems the Earth is facing now. Colourcash is a way to deal with those problems. The principles described in the Earth Chapter about care and responsibility for everything alive, are "pretty" and "nice".

We leave it up to the individual colourcash participant to sign the Earth Charter. This is to be decided amongst the participants. It's is not my place, as the author of this book, to bind you to a Charter, nice as it may be.

Energy

This word is, after 'the' and 'and', probably the most used word in this book. Read the entire book to get the point. When that takes to much energy, you realize that too much energy isn't good either. Too little energy is also an imbalance, therefore I added this entry.

Excise Duty

Excise duty is a special tax on consumption goods that are legal, but their use is discouraged. E.g. Alcohol and tobacco, but also petrol. Let's see what duty is needed in principle for each of these products before we ax this unpopular notion.

Alcohol becomes a lot more expensive (see Alcohol); also because of the water used to produce lets say beer, the high cost in transport and the distance of the sources of raw materials. When there is a possibility to produce a local alcoholic beverage, the price will be so high that an extra duty won't be needed.

Fuels are a lot more extreme, even though just a little more expensive (in Red) per litre, the consumption lies a lot higher. Compared to the Euro price, petrol will be about ten times as expensive. No need for excise duty there.

Tobacco is addictive and is unhealthy, and it's imported. I predict we will pay only Red and Blue. Taking the content of the ashtrays back to the plantations I don't see feasible. This will convince producers and users alike to change to other, more environmentally friendly alternatives such as Sage, Lungwort, Coltsfoot, Rib-grass, Peppermint and Red Clover. Also Hemp can be taken into consideration. In fact we will return to a pre 1600 situation when tobacco was still unknown in Holland and everybody smoked home grown plants. We'll have to wait and see whether we need excise duty for these alternatives, and this will depend on their addictive effects, effects on health and price.

Frequently Asked Questions

Is this a joke?

That's up to you to determine. colourcash demands the simultaneous use of both left and right hand side of the brain. If you can only associate this with a joke, this says more about you than about colourcash.

Is this a new kind of Communism?

No. The section on Communism explains this in detail. See the section on page 114.

How can I join?

I am the author of this book, but at this point, I am no more than the author. Please use your own network and get your contacts involved to get this thing off the ground.

Is one allowed to emit one's own money?

The Dutch State dit not block the Raam, the money published by Maharishi Mahesh Yogi in Limburg. So, yes, you can publish your own money. More important colourcash is not money! Money doesn't mean anything and colourcash does.

There is no "transfer of ownership", only "transfer of responsibility. The colourcash is only there to register this responsibility. There is no "aim for profit" or interest. There is no money. The State has nothing to do with it.

By the way, the example of the Mahirishi is just that, an example. It is not within our goals to set up a commune, sect or otherwise closed group.

Is one allowed to set up one's own community?

VN-resolution 1514.2 says that anybody has the right to develop socially, cultural and economically.

What are the advantages for me?

It's important for you to reach balance, equilibrium. Whether you reach this by getting extra advantages or by cleaning up a number of disadvantages in your own day-to-day life, is up to you. It's up to the individual to find their own reasons to join colourcash.

What has all this to do with God?

That is for you to decide.

How can I still get product X?

There are three possibilities. You could buy it for Euro on the Black and white market, no changes there. But thanks to colourcash repairing things has gotten a lot cheaper. Why not trying to get a broken appliance that you can use for spare parts? There is also the possibility to "liberate" technology. We need a lot of tools for e.g. butter, cheese, textile and rope that we can get from museums. Modern technology is often patented or uses industrial processes. Industrial machinery could be part of a colourcash inventory but will be one of the last things you really need.

How can I get service Y?

Services are lot easier to be had than products. In essence they are Man-hours. However, we depend on external companies for services as telephone, rail roads, electronics, internet, gas and water.

Maybe we should take a good look at this infrastructure to re-invent it in a better, free-er way. Not a rail road but a horse tram for example. A local network in your neighbourhood with a connection to a commercial provider. Your own electricity using solar panels and windmills. DIY is be far the best solution here.

Export and Import

International trade is an important means for making profit in the modern Black and White economy. But the commerce grew out of bounds. Soup you can make better and tastier yourself, but it comes in a tin can from 30 countries. We don't produce any more clothes in Holland, everything is imported from Asia. In other words; there is no balance between import-export and own production for domestic use.

International trade is still possible under colourcash. Trade as much as you want. What's important is that you make agreements about which colour to use for payments, and where it's valid. It's important in colourcash to keep the money in your "own" cycle. When all the money flows to one side, the other side runs out of money. We already know this from the Black and White economy. Colourcash has a built-in security mechanism to avoid this kind of money-flight.

The same mechanism makes it so that you can't import or export limitlessly from one side only. Countries can't be sucked dry any more, as is the case now. Also for goods the circle has to close, unless you want to live on a Blue heap or turn into a Smurf.

Free Market

The most important dogma of the market-fundamentalists is calling the market a free market. There is no "obligation" in your choice of clients or suppliers, nor are there fixed prices so the market is "free".

The market in itself is rather free, that is correct, but the door to this market is secured with barbed wire and heavily guarded. Every entrepreneur has to comply with an encyclopaedia of rules; all ordained by institutions whose names have been abbreviated to three or four letter acronyms. And when you're business is food, then it really gets complicated. When you have a croissantery, make sure the space in front of the oven doesn't exceed 19 degrees centigrade. That will cost you a fortune in air-conditioning. Only affordable for big chains that can't wait to take over your business after you were forced into closure. The mill in Maassluis turns, but doesn't mill, the stones didn't pass the last test. They still work fine, but the miller is fed up. I'd better stop complaining because you could fill a book with the phenomenon of over-regulation.

Consider this dogma:

> *The Free market system is the best. When you free something, it works automatically in the best possible way.*

The beauty of the notion "Free Market" is that you can let go of the "Mother of all markets". The introduction of an alternative money system is nothing more than a competitor to the Capitalist system. A taste of its own medicine.

> *The Free Market is only really Free when other market systems are welcome on the "Mother of all markets", so that suppliers, traders, manufacturers and consumers can choose for the best, the most suitable, alternative.*

We will see whether providers of financial services, such as banks and insurers want to make use of this available space.

When banks and insurers don't jump to this opportunity, users of colourcash will fill the gaps with their own services.

And it's the same for other professions, services and products. It's up to the individual to choose for the most suitable system. I don't mean to force colourcash as only and best system. Let's use it as a complimentary system for the Euro system for now.

When a government prohibits colourcash, it limits "Free Market" to the Capitalist system, and acts against her own principles.

Fuel

We are used to low fuel prices. Petrol costs between Euro 1 and 1.50 per litre. A kilo of cauliflower costs a bit more. This is going to change. The Green value of a kilo of cauliflower and a litre of petrol are both 10 Green. 1 Green per 100 grams. Stays more or less the same as in the Capitalist system.

But here's the trick. We're going to take the Red value of the products. Cauliflower is not as expensive. It's in the lowest class, 1/8 of the index, the index being sunflower oil, 8/8 or 1. Petrol contains more energy than sunflower oil. Powerful stuff. So on the scale it goes to class 10, so 10/8. Petrol will be about ten times more expensive then cauliflower.

Since the exhausts are dumped in the air, petrol has a Blue price in the ledger. A shame. Buying petrol is very expensive in Red and Green. You can earn Red through labour, so you can make up for that. OK; you'd have to save between 50% and 70%, but that's easily done. But you can't recover Blue. Buying Green and activating it as Blue, that's going to show. Your financial data are public, so you are a marked man. You could borrow for a while, but then you'd have to reconsider your lifestyle. Maybe a sailing boat or a horse after all...

Greenhouse effect

The Greenhouse effect is one of the many "side effects" of the economy we know today. It can't be helped; we can't stop the huge amount of greenhouse gases emitted daily. Measures such as a huge ecotax are juridical and politically impossible. This is the result of our urge to compete. We are all so busy competing that we're screwing up the Earth.

Fuel combustion wouldn't be all that dramatic if we could recycle the CO_2 released. It doesn't matter whether it's fuel from sunflower oil, corn, colza, the air would smell of fried potatoes. When you have enough space to grow your own fuel and keep doing so every year, there's no problem.

But there is a problem with the present scale of oil use. I'm not saying that oil use is good or bad. I'm not saying bio-energy is good or bad. It's about balance. We've lost the balance. Our use of fossil fuels is gigantic and there is hardly any renewal. This is bound to get us into trouble. The problem is the "cycle", as always. In this case the exhausts stay in the atmosphere. These gases are mainly made up of water vapour and carbon dioxide (CO_2). I never hear anybody complain about the water vapour, but the CO_2 is a problem that leads to climate changes according to climatologists. Scientists, indeed.

What is happening on a global scale with the climate you can see in colourcash in the accounting. You will notice when a country has to import a lot of oil. You pay Red and Green for oil, and you don't get anything in return. After use Red is burned and gone.

Interest

Interest is prohibited in colourcash. Just for argument's sake I will explain what reasons Capitalists use to raise interest. They assume there is growth (I'll come back to this in a minute). With this growth you can invest your money in a profitable way... A typical example is the tree, because it grows and after x amount of time, the tree has grown y number of centimetres. Because it's an apple tree, you will get always more and more apples and so you will have a profit. Money lenders lend you the money, so they couldn't buy trees themselves. They missed the growth of the tree, and that is _your_ fault, because borrowed the money. There you have it; interest is compensation for a missed income generated by property. Because there is nothing but growth.

But they miss one important thing: there is such a thing as wood rot. Worms plagues. Besides growth there is also decay. Death.

They had forgotten this for a minute. A collective brain malfunction you could call it. When a tree dies, it doesn't produce any apples and more and so the investment is lost-This could happen to anyone. But not to the money lender, he has a guaranteed tariff of 7.6% interest per annum, whether the tree lives or dies. And when the tree dies, the borrower goes bust and has to leave his house, out in the street to give the lender some of his money back. The tree is cut down by the next investor to make space for a cooling cell to store Chilean apples...

I'm not saying interest is unreasonable, it's irrational. Based on deceit. A big lie. That is why I refuse to borrow and I should stop lending... Shall I take all my money out of the bank?

The last tenner

The story of the last tenner is famous. It shows how money is created and how we get our sky high, never to be paid back debts.

Imagine a man shows up in a village where there is no money. He lends ten Euro to ten men. But, as interest, each of the men has to pay back 10% interest, so 11 Euro.

After a year 9 of the 10 men are able to repay those 11 Euro. But the tenth man only has one Euro. He's a tenner short! Guess how that's possible?

There were only 10 x 10 = 100 Euro emitted in the village. The ten men have to pay those back, plus interest. Interest at 10% over 100 Euro is 10 Euro. In total 110 Euro have to be paid back. However there were only 100 Euro in the village to begin with!

The last 10 Euro are the cause of the first bankruptcy. Say that the lender generously allows the village to pay back a year later. But then you pay 10% interest on 110 Euro, because that is the amount owed after 1 year. Can you predict if the man will have the money year later?

Labour

Labour, work, pay, salary, wages... Of course you mean the work you do, but the words also indicate retribution. Nowadays hourly rate is popular, other than that we also know piece work. Minimum base wage is not very common.

In colourcash I presume this will change. Minimum base wage, guaranteed by your local community is on top of the list, followed by piece work. Hourly rate will become less important, due to the proposition *Time Is Art,* or, as others say, *In Time Is The Art.* This means that real value is in doing the work on the right moment, instead of counting all the unused, wasted time. This holds for creative people and agriculture.

Because larger life cycles are more expensive than smaller, short life cycles, the trend towards globalisation and up-scaling will be bent towards downscaling and localisation. This means that there will be less demand for boring, monotonous labour. Work close to home will be required for most jobs, because fuel will be so expensive. That is fun, you'll see your children's faces again instead of the dashboard of your car.

Labour disability

Unsuitable labour, that is what people are "disabled" for. Unsuitable labour is the only cause of labour disability we can remove. Here's how:

Colourcash is a "game" which is in search for balance. It's of no use whatsoever to squeeze the maximum out of workers. The "maximum" reveals itself later as a deficit. The string can't always be strung tight. Balance. When you find balance, you don't go from extreme to extreme, like a drunken ballet dancer

Livestock

At the moment the Dutch have millions of animals. 150 Million chickens, 20 million cows, you name it. Because hardly any one is a cattle or fowl farmer in the traditional sense of the word, these animals live in "batteries".

The impressive amounts of feed we buy abroad. The manure

surplus hasn't been solved completely, but the sting has been taken out of it for now. The milk quote and various plagues like BSE, foot and mouth-disease, aviary flu and swine fever wreaked havoc on the industry. Over 50%of the Dutch people turned, at least part time, vegetarian.

In colourcash it is advantageous to keep only a small amount of animals. I see an animal-friendly future, combined with some small-scale farming. Eating meat will cost about 7x the amount of Red than food from plants. Also milk, cheese and eggs will be more expensive, relatively speaking. On a small scale there is no problem to recycle manure, it can be used on the land, so the cycle is closed and there is no Green on the Blue heap.

Small scale animal keeping becomes economically viable in colourcash, whilst it is madness to farm animals on a small scale in a Black and White economy.

Medicine

Modern health industry is a combination of petro chemistry and electro techniques. Old-fashioned hands-on practice (like checking a pulse) is a lost art. Impossible for an MD to do this in 7 minutes per patient.

The more patients a doctor have, the higher his income. This isn't logical at all. One would say: the more sick people, the lower the income. This motivates people to get well. Taking more pills to counteract the side effects of other pills is a serpent biting its tail. This serpent is inherent to the system of research and financing in healthcare in general.

Century old and new healing sciences are rejected by "science" because they are outside their thinking box. But, healthcare the way we know it now rose in the era of Classical Mechanics. The laws of Action and Reaction are central in this way of thinking. Since Quantum mechanics expanded this frame of reference, health care hasn't changed yet. Modernization of our healthcare system is delayed until its feeding ground, the financial system, is changed. This is going to happen now.

More, see Drugs and Medication on page 121.

Organic

In Dutch, *organic* is called b*iologisch*. There is a TV-campaign saying "Biological, it's so logical".

But more expensive, says the shopkeeper. But healthier adds the doctor. But animal friendly, say the environmentalists. Which of the above we can delete? Let's delete the text of the shopkeeper. colourcash is biological - organic - and therefore always Logical. And cheaper, and healthier, and better for the animals. There can only be One who's the most balanced.

To buy chemical fertilizer you will have to pay a lot of Red and Blue. The profit doesn't rise however. Nor does the nutritional value. It's possible the land yields more kilos of produce, but that is mostly water. The taste gets diluted.

The higher Red and Blue costs are immediately reflected in the price. Food should only cost Red and Green. Every hint of Blue immediately is noted. Hey! That guy still buys poison!

Peak Oil

In Shell's annual report, there is a subtle mentioning of a

mayor fall in oil distribution. 10752 barrels per day in 2005, against 12760 barrels per day in 2004. This is a drop of 15%. You couldn't tell by looking at the profit figures, they kept growing.

The American oil production already declined in the 1970's. The downward trend in the US was about 35 years before the global downward trend in oil production.

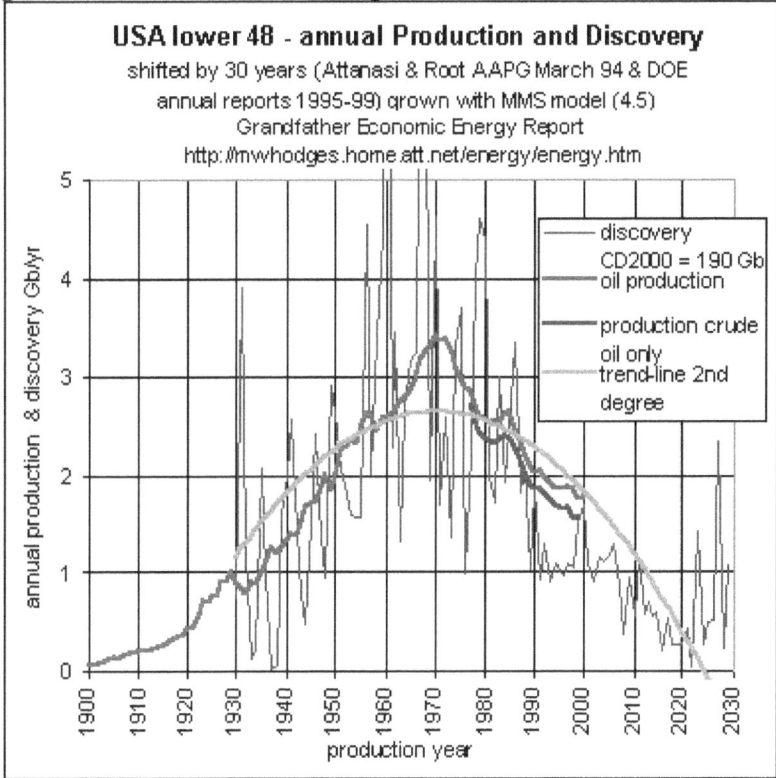

The American oil production already declined in the 1970's. The downward trend in the US was about 35 years before the global downward trend in oil production.

In May 2000, a famous Dutch politician Frits Bolkestein wrote in an open letter to the newspaper "De Volkskrant" about the underlying phenomenon: Peak Oil. What it boils down to is that oil production has peaked. The pressure has gone and now the oil doesn't come to the surface as easily as before. The oil hasn't been used up yet, not by a long shot, we are exactly halfway through it. It's generally assumed the peak was in 2006 or 2007. For Shell it seams this peak was in 2004.

The same thing happened in the 70's in the US oil production. The oil wells aren't empty yet, but they yield less oil. New sources are found, but finding them and exploiting them, yields less than the reduction in production in other locations. We are talking about tens of thousands of wells that are producing less. Those few new big ones don't really help.

What are the consequences of this peak? First, the media, the general, commercial media that is, doesn't report on it, so the big companies keep this info to themselves. The goal Shell put itself is, when you read the annual report closely, to conquer the "alternative fuel market", so they have a business to run in the long term as well. I haven't seen any reaction from the side of the Dutch State. They just keep building motorways.

Oil prices are going up. A lot of people think petrol is expensive at Euro 1.30 per litre. It depends on your frame of reference, I tell them, just to get them going, that petrol will become 10 to 30 times more expensive than cauliflower. This is the end point where the price becomes stable. In other words, the balance won't be reached below that ratio.

The consequence is that using your car to go to work becomes unaffordable. 30 or 40 km alone in your car, commuting? You'd better stay at home, you would make more. People will car pool, sleep at the company twice a week, change jobs or places of residence and eventually start using (horizontal) bicycles.

Also heating your house will cost a fortune. Start taking into account that a big expensive house looks nice, but an economic house is the house of the future. I remember the castles where it was great living in summer, but freezing cold in winter. The nobles lived in towns in winter, where they were close to a warm tavern. When you live on a farm, there is nothing to worry about. The cows emit enough heat that sleeping on the deal is luxury.

Pension

Your pension, if you are to have one, has been safely invested in big pension funds. Let's see where it was invested... State bonds. Real estate, or in other words offices and residences. I can't speculate on the future. I won't make any predictions about how these investments will fare.

I do know how much pension I've saved up so far. In 2036 I will get a "royal" 900 Euro per year. Whenever a pension advisor comes to me I ask him how many bags of potatoes that will buy me in 2036. The advisor always tells me that nobody knows, because nobody knows what money will be worth in 2036.

With this one question I put them in their proper place. I don't care how much money my pension will be worth. I want to live in a house, eat every day and have decent health care. How much is that going to cost me in 2036?

And that is the big colourcash advantage. Prices are fixed. In so far that when you get your potatoes from far away, they will be more expensive. But those are real price differences. No speculation, no hyperinflation.

If you want to spend the twilight years of your life, like me, living in a colourcash system, we must start now. No-one else will take care of it for us. This is what is called taking up your responsibilities these days. Very modern. My pension? A rocking chair, a French farm, all kinds of animals and a house filled with family.

Permits

If you ever have tried to set up your own business, you know how hard it is to get the necessary permits. There are soooo many rules you have to comply with. You have to be in the right category. When you're an outsider, it soon gets difficult, not to say impossible. Small entrepreneurs are countered on a grand scale because of permits. The Big Boys don't seem to have any problems with all those permits. Especially environmental permits seem to be handed out like candy. And once the permits are obtained, there is no more control. There can't be any more control, because the services in charge of controlling are way too small. Overtaxed and understaffed, they themselves are bound hand and feet by too many rules. The centralist government organizations are only able to go after the little guy, to make him close down his business. They won't tackle the big companies, besides giving them a ridiculously small fine.

In colourcash we turn, you've guessed it, the whole thing upside down. There is no central government that decides who can and who can't. You can do anything, but you are and remain responsible. And everything is seen by everyone. You can no longer dump some gas, at night when there is a strong wind blowing. This "missing gas" immediately shows up in the books.

Emission and control of permits are decentralized. The ones who decide whether company X can continue their activities in the economic cycle are the suppliers and the others in that cycle. There are millions of eyes and ears and they can all take a look at how business is run. No central government that doesn't seem up to this task anyway, but a transparent, public network of people who work together. Together they grant you your permit! One case of "Brent Spar" and you're out of business!

Pollution

Pollution is a subject that was on the original agenda to be tackled by the fore runners of colourcash. The essence of pollution is that you can afford to throw away stuff. Unnoticed, unseen and, in the least, unpunished.

Standard example is someone who drives a car. You buy a tank full of petrol, 50liters for instance. You pay. 65 Euro. After two weeks your tank runs dry. The fuel was burned and blown into the atmosphere via the exhaust pipe. To burn petrol you sent in oxygen via the air intake. That oxygen is now linked to carbon from the petrol. This we call CO_2. Then there is whole range of other poisonous gases that were released. We also released water vapour, the all too visible smoke.

Now there is well over 50 kilos of CO_2 in the atmosphere and this still belongs to the driver of the car. Did he sell it? No. Did someone say they would take care of it? No. Morally speaking the CO_2 that was released is still the responsibility of the person who bought the petrol in the first place. Only now drivers and car owners think it normal to just dump their co_2 in the air. And anyway, in a Euro economy nobody charges you for it. Gone is gone, and nobody is responsible. And anyhow, everybody does it, don't they?

colourcash doesn't work this way. You buy 50 litres of petrol for about R 14 G1 B0 (per 100 gram). A total of R700 G500 B0. And now you drive. Gone is gone. The beauty of it is that you can recover the amount of Red through labour, but those Green 500 are gone forever. This will show up on the end balance. You remain responsible, even though you think the CO_2 has disappeared.

Profit

What do you need Profit for? Is a normal, stable income not enough?

When I read the stock market pages in the newspaper, profit in itself isn't enough either. The game is about profit growth. Every year your profit has to go up by say 10%, otherwise you should choose another investment.

Profit is the motif of people who only care about the result, get the result, harvest so to speak, and leave the next guy with the mess. I see profit as an extra, a bonus, a piece of "imbalance". Soon enough this profit will be compensated by a loss. Am I in balance on a longer term?

Short term "profit hunting" is mainly caused because there is such a thing as interest. This is rather complicated, but most companies don't care if long term investments don't yield any results. Governments don't plan long term either. Otherwise they would have dealt with pollution a long time ago.

The main goal is to get profit now. When something will give you 1 000 000 in profit in 2099, it isn't interesting. Because of interest the cash value of "money" that I deposit now in the bank to get that money later is practically nil. You only have to put Euro 4430 in a bank account (2006 6% interest) to have 1000 000 in 2099. Nobody cares about that.

It's essential that profit is about money. Logical? No, because I myself care more about living in a healthy house, good food and a comfortable life. You can't eat profit.

Taxation

Taxes are the motor of every power. The point is, the Ones in Power have been replaced, together with capitalism by a community. The community shares among itself, no need for a central "power that commands". So there is no need for taxes, there is nobody to collect it!

Say you would collect taxes to pay for collective services, what would it look like in colourcash? When one of the colours Green or Blue is taxed, the tax service (?) breaks the cycle and this leads to more Blue. So we're left with the option to tax Red. What does a Red tax look like?

Labour could be taxed. The more you work, the more you earn. As punishment you're to pay.

It makes more sense to tax "Space". The more Space you use, the more Sun you could have captured, the higher your income could have been. You pay for this. When you run a Bonsai-nursery, small but labour intensive, you pay less tax than someone who has a field with cows. The owners of productive land share in the profits anyway (what else are you to do with tons of corn cobs?) I don't know what is to be taxed there.

During the Change over it is possible the Dutch State or her Tax Service comes knocking bearing a "blue letter", asking us to pay tax on our turnover, our income. And we'd have to pay Euro for our colours.

Just like the LETS systems, we could propose a maximum turnover. I know that in Utrecht the limit was 3000 stars. The Tax Bureau still has this rule on their website.

Essential here is that the Stars were linked to the Guilder. One star equalled 1 Guilder. In colourcash things are more complicated. There is no connection between the Euro and Colourcash. As a result you can't simply say; 3000 Red is the tax free ceiling because all rapports are different.

It gets even worse when you take into account that colourcash, being a closed system, is troubled by the State. It's because of the State there are such huge amounts of toxic substances on our fields. All Blue. The water, even the own well, often is polluted. All Blue. Shall we settle our health bills caused by the State by building a freeway next to our house so our kids all have asthma and we all suffer from a chronic frontal sinusitis?

No, the Tax Service will say, because we don't reckon in those weird Blue-units of yours. Neither do we reckon in those strange Euro, we say than. Good day to you!

State

The State only justifies its existence by the Money it emits. In that sense the Sovereign Dutch State has signed its own death penalty by changing the Guilder with the Euro. You can also see that its power gets less and less by the day, in favour of EU, NATO, WTO etc.

By not recognizing the Currency of the State, I run the risk of being arrested as Enemy of the State. And rightfully so. There is no greater enemy than the one that claims that the money the State emits is based on an illusion and a lie, and presents a better plan to boot. The question remains whether it has the judicial means to arrest me. The Question whether the people will vote with its feet is even bigger. (Translator's note, Voeten, the author's name is translated as Feet...)

It's all a matter of transition period. You'll see that we can get by without governments just fine. They are made up of politicians, and those are just as insecure as their heads are big, and civil servants (the ones who follow orders).

The most important government task is shifting money around. At least, that's all politics do, leaving aside symbol legislation. I don't really know what else they do, but all the rules they've cooked up for schools, doctors, hospitals etc. we can do without. Frankly, I don't even know all these rules. A complete Judicial Code book costs thousands Euro, is about three feet thick and can only be read by lawyers. Do you think that is of any use to us? It is of use to the ones who claim "property", the ones who consider themselves "rich". The story came to an end.

From the grass roots up we will find new rules, hopefully we won't need many of those. Grass roots means there is no "Government", but more a "Sovereign entity". I am my own Sovereign Entity. So that makes the Sovereign Entity of I, "my" girlfriend and the plans we have in our house. We don't have any animals yet, but they would have been included. The next SE (Sovereign Entity) is with the neighbours, the building, the street, whatever. You make up your own.

Remember the rules of section on Law and Politics on page 74? *Nobody can force their will to someone else. A majority does not overrule a minority. Only when everybody agrees a decision is taken and before that… the time isn't ripe yet.*

Statistics

An important tool for managing a country is statistics. In fact it's a technique of numbers. This is the place to manipulate if you want to. How many unemployed people are there? That depends on the way you count them. A lot of unemployed people were put into welfare and are not included. How many people with work disability are there? That depends on the way you count them. A lot of disabled people are put into welfare and don't count... How many people are there on the dole? Oh, we can get that info in a second!

The same goes for "the economy". You know, our economic growth. President George W. Bush had to deal with a falling economy. He pumped over 70 billion dollars into the war with Iraq. Nice boost. All of a sudden the economy is attractive again. There you go!

When you listen to Business Radio, you hear all the numbers in percentages. The percent. All measures in nature have a unit, the way the litre and liquids go together. A percent doesn't have that. The measure of a percent stands alone. A percent has no meaning by definition. A good reason not to use it. A unit of measure without content is the best guarantee for an empty... yes, content!

Subsidies, grants

Subsidies are one of the ways the State uses to shift money. Especially in situations where the market mechanism isn't foolproof, subsidies and grants are used. Education about nature and environment for example. You can't make any money from these, but you do need it. Time to open the grant-valves.

This system will be changed by the common decision to do a determined task. Awareness and consciousness take over the task subsidies are supposed to do now. The inner mechanisms of colourcash do the rest. When people get the opportunity to give meaning to their lives, most of the time they will. The others need help, not grants.

Traffic and Transport

You will have noticed we live in a world where the car plays an important role. When you, like me, don't drive a car you probably still depend on it. I still do my shopping at the supermarket and the greengrocers. What will happen if their lorries stop driving? What will happen when you have to pay the real value of petrol?

I calculated how much energy petrol contains and then I compared it with the amount of work I do in a day. It's a tragedy, but because a car is so heavy and drives so fast, you often use more energy in 10 or 20 kilometres than you can recover in a day. Or, you might as well not go to your place of work, you earn more. Even more advantageous is moving or looking for another job.

I am not going to speculate about future prices of petrol, not about the Euro price, nor in RGB, the colourcash currency. I just state that there is about 30 times more energy in a kilo of petrol than there is in a kilo of cauliflower. Whether that means petrol will be 30 times as expensive as cauliflower, who knows? The fact remains that it's cheaper to get a bicycle. 30 kilos of cauliflower give me enough energy to pedal for about two weeks. And that gets me to France. That one kilo of petrol won't even get me to Antwerp!

Unemployment

There is plenty of work. But a lot of it is done by machines. Before robots and computers took over the work, almost everybody was a farmer to some extent. Then factories came, where fabrics were made. Around these factories cities rose up. They became home to farmers who now had become factory workers. Nowadays, these factory workers work in an office and live in "sleep cities" and suburbs.

These changes were made possible under the Euro and Dollar system. Its days are numbered. In the future we determine the economical order. We allow ourselves the choice to work where we want and only use machines if we don't want to do the work ourselves.

VAT

Value Added Tax or VAT is a tax when companies buy and sell products. A better name would be PAT, Price Added Tax. A fixed percentage of the price has to be added to this price and paid to the government. When the product becomes more expensive when it's sold again, the entrepreneur has to pay that percentage in taxes to the government. The rest of the amount he pays through his supplier. The whole process is controlled with invoices and VAT numbers.

In essence VAT is based on the economic production chain, the sale of a product from the cradle to the grave, from raw material to consumer. In the end the consumer pays the tax on the value of the product, or rather the price of the product. In colourcash you're dealing with a cycle of companies that construct and destruct the product together. So, what's the deal in this cycle?

You could introduce a similar tax on the colour Red. Nice and pretty. As long as the product grows, it gets taxed, but how to deal with it in the second part of the cycle, that remains the question.

Say you have a product that can be reused after use, before it has to be destroyed forever. You could divide the tax over a number of users. You use the same principle as VAT does, but the other way round. One consumer sells it at a lower Red value; the difference is what he used up. He can charge a bit of VAT for this, and this will compensate for the part of the value on which he paid taxes but didn't use.

Whether it makes sense to levy taxes like this in the colourcash system remains to be seen.

Water

Water is essential for our life form. Fresh water. We use an awful lot of it every day. You could say we use too much. But that's not the point here. The point is, what do we do with it after use? And that's where it all goes wrong.

All too often water, after it's been used only once, is dumped and disposed of. In principle it goes to the sea via canals. Exit water. Not dramatic for river water, because that was going towards the sea anyway. But it is dramatic for water we pumped up form the ground, precious water reserves that took thousands of years to be built up by nature.

Of course we think "recycle" also here, as in every part of colourcash: recycle recycle recycle. And then we try to keep the cycle as short as possible. For water this means re-use, re-use re-use. The smaller the scale, the better it is.

Water catching becomes very attractive in colourcash. Gray water circuits are very economical and earn back their costs sooner than in a Black and White economy. This is obviously because prices are set up differently. Recycling is the essential criterion.

The water company pumps up our water from subterranean reserves. Long ago this water seeped in. You can't pump up more water from below the earth than is stored in the first place. When you pump up more than is added by seepage, the balance gets disturbed. That water no longer is Green but Blue. Depending on the Blue stream this is a very disadvantageous situation, because this blue supply can run out any time. Or, the Blue supply should be stopped as soon as possible.

The other supplies, via rivers and rain water, are Green. Income. Nice! Now the trick is to keep this water Green as long as possible. Recycle, recycle, and recycle. Dumping is possible, but whereto? When the water stays in your own cycle, no problem. But when you take it elsewhere, it turned Blue.

Everybody knows bottled sparkling water. Mineral water. Very often it's pumped up at far away sources and put into boxes or bottles, then transported by lorries to the corner shop. 500 kilometres is a normal travel distance for a bottle of water.

What do you think? Is this water still in its original cycle? I don't think so. If nothing went wrong at the pumping up, then surely the transport messed things up royally.

An important or rather THE most important water guzzler is agriculture. How else can a cucumber stay juicy if not with water? Often a lot of water is lost by the indiscriminate spraying of water on land that doesn't seem to have a saturation point. A little shower does more good to crops than a water cannon spouting water all day long.

When you interrupt the water cycle by irrigating the land by spraying, keep in mind that breaking this cycle also has consequences for the surroundings. The vaporized water has to go somewhere. It rises through heat. And rising air doesn't permit rain to fall. The best way to chase away the rain is to make it rain artificially. Method and effect together are a quick way to get rid of all our fresh water reserves! One day Lake Aral was the biggest lake in the world. Now it's a barren desert, surrounded by barren cotton plantations... If the government of the Soviet Union had learned their lesson sooner, there would still be a lake and there would still be cotton growers!

The future is called permaculture, the hands-on counterpart of colourcash, with a variety of techniques enabling water-less gardening. Water-collection is a simple way to have at least some free water. A different selection of plants or crops will make a lot of impact. Finally, mulching keeps the soil moist and cool for a long time. Good luck!

Without money

More and more people are beginning to realize that colourcash isn't the last step in idealism. They are looking for a world without money. Saves a lot of hassle. Being conscious and behaving consciously is what matters, whether and how you register this is a lot less important, if you know what you are doing at least.

And there is the knot. The whole world at this moment (2009) is off balance. More and more material is destroyed or burned, not recycle-able. Water supplies are under pressure, an oil peak is approaching rapidly. More and more people have to rely on the food bank, or, in Black and White, mankind and planet are doing worse and worse.

So we have to make sure we recycle everything, limit the use of energy, learn how to co-operate, share more, laugh together instead of being afraid of each other...

Once we've learned all this and we are in balance, and that could take a couple of centuries, we can put the RGB currency in a museum, next to the Pound, the dollar and the Euro.

Woman

For the last 2000 years, probably a lot longer, women have been dominated by men. Women couldn't be priests, couldn't vote. Women were only "women" and not always counted as human beings. John Lennon once wrote: "Woman is the nigger of the world".

When I wrote this book politics was a male dominated game. The few women we see in politics carry the same hard balls as men do. No caring, loving, unifying and sharing in politics.

The "female, caring way has "snowed under". This snow will melt. We will see that it was only our window that was covered in snow. Once you've found the coloured window, you will see the value of the caring "female" aspects of life.

In real life this means that we won't wait for men to say yes to colourcash. We will help women with colourcash. This is THEIR thing!

Literature

Willem Hoogendijk, *Economie Ondersteboven*. Utrecht, Jan van Arkel, 1993.

Vereniging voor Ecologische Leef- en Teeltwijze (VELT), *Handboek Ecologisch Tuinieren*, Berchem (België) VELT, 9e druk, 2002.

Bernard Lietaer, *Het geld van de Toekomst*, Amsterdam, de Boekerij, 2001.

D.H.Meadows, D.L.Meadows & J. Randers, *Beyond the limits. Global collapse or sustainable future*, London, Earth Scan Publ. Ltd, 1992.

Peter Toonen, *2012, Het einde van de gestolen tijd. Het verhaal van de Maya´s, de bewaarders van de universele tijd*, Andromeda, 2005

Rob Hopkins, *The Transition Town Handbook*. Green Books. 2008.

Toby Hemenway, Gaia's Garden, A guide to Home-Scale permaculture. Chelsea Green Publishing Company, 2000

www.ingramcontent.com/pod-product-compliance
Lightning Source LLC
Chambersburg PA
CBHW031941190326
41519CB00007B/610